Feminism & Science

RACE, GENDER, AND SCIENCE

Anne Fausto-Sterling, General Editor

Feminism & Science

Edited by

Nancy Tuana

Indiana University Press
Bloomington and Indianapolis
1989

Manufactured in the United States of America

Library of Congress Cataloging-in-Publication Data

Feminism and science.

(Race, gender, and science series)

Bibliography: p.

Includes index.

1. Sexism in science. 2. Feminism. I. Tuana, Nancy.
II. Series.

Q175.5.F45 1989 305.4'35 88–46044

ISBN 0–253–36045–5

ISBN 0–253–20525–5 (pbk.)

1 2 3 4 5 93 92 91 90 89

CONTENTS

PREFACE / vii

ACKNOWLEDGMENTS / xiii

OVERVIEW

Feminist Scholarship in the Sciences: Where Are We Now and When Can We Expect a Theoretical Breakthrough? 3
Sue V. Rosser

FEMINIST THEORIES OF SCIENCE

Is There a Feminist Method? 17
Sandra Harding

The Gender/Science System: or, Is Sex to Gender as Nature Is to Science? 33
Evelyn Fox Keller

Can There Be a Feminist Science? 45
Helen E. Longino

Is the Subject of Science Sexed? 58
Luce Irigaray / translated by Carol Mastrangelo Bové

Uncovering Gynocentric Science 69
Ruth Ginzberg

Justifying Feminist Social Science 85
Linda Alcoff

John Dewey and Evelyn Fox Keller: A Shared Epistemological Tradition 104
Lisa Heldke

FEMINIST CRITIQUES OF THE PRACTICE OF SCIENCE

Science, Facts, and Feminism 119
Ruth Hubbard

Modeling the Gender Politics in Science 132
Elizabeth Potter

The Weaker Seed: The Sexist Bias of Reproductive Theory 147
Nancy Tuana

The Importance of Feminist Critique for Contemporary Cell Biology 172
The Biology and Gender Study Group

The Premenstrual Syndrome: "Dis-easing" the Female Cycle 188
Jacquelyn N. Zita

Women and the Mismeasure of Thought 211
Judith Genova

BIBLIOGRAPHY / 229
NOTES ON CONTRIBUTORS / 240
INDEX / 243

Preface

A being like the female, without the power of making concepts, is unable to make judgments. In her "mind" subjective and objective are not separated; there is no possibility of making judgments, and no possibility of reaching, or of desiring, truth. No woman is really interested in science; she may deceive herself and many good men, but bad psychologists, by thinking so. (Weininger 1906, 194)

The politics of science have included woman within the gaze of science but have excluded her from the practice of science. Weininger, in developing his "science of character," believed himself fully capable of describing and accounting for the nature of woman. In doing so, he defined women out of science. To be a scientist one must be objective—woman is incapable of objectivity. A scientist makes rational judgments—woman is incapable of reason. A scientist desires truth—woman desires only truth's opposite, passion.

This conception of woman's nature has excluded us from the very process of defining ourselves. Our silence is dictated; we are made into an object of study. But this exclusion has also precluded our participation in the defining of science. Through the feminist critiques of science developed over the last decade, we have given voice to our views about the politics and practice of science. We have uncovered the complex interconnection of sexist, racist, and classist biases grounding theories of human nature, and in doing so have seen the ways in which such biases permeate the entire structure of science. Feminist critiques of science have thus begun to focus on the ideologies, politics, epistemologies, economies, and metaphysics of traditional science.

This analysis has given rise to a number of difficult questions. "What kind/kinds of science is/are consistent with our critiques?" "Are we asking for a feminist science?" "Is there a feminist method?" "Why should feminist values take precedent over others?" "Do women do science differently than men?" "Is sexist science bad science?"

In recognition of the importance of these topics for feminist scholars and for philosophers of science, two issues of *Hypatia: A Journal of Feminist Philosophy* were devoted to the topic of feminism and science [1987 2(3), 1988 3(1)]. The essays reprinted here have been selected from those special issues of *Hypatia*. These essays offer a variety of perspectives. Many address the above questions and the controversies raised by them. Others trace the role of gender politics in the practice of science. Some attempt to offer future visions of a practice of science consistent with feminist values. Through them we see the variety of perspectives, epistemologies, and values involved in an examination of "the science question in feminism."

Sue V. Rosser opens the dialogue with an overview of feminist scholarship on women and science. Her work provides a helpful synthesis for readers familiar with the literature, and an introduction for those new to this area of feminist research. Her discussion of feminist critiques of science, feminine science, and feminist theories of science, is particularly relevant in that the articles in this anthology fall within one or more of these categories.

I have employed two of these categories to organize the articles in this anthology: feminist critiques of the practice of science and feminist theories of science. The first group of articles focuses primarily upon feminist analyses of the epistemological frameworks of modern science, or what Rosser calls feminist theories of science. The second group of articles involves critiques of the ways in which the practice of science is affected by, and in turn reinforces, sexist, as well as racist, homophobic, and classist, biases. This division, however, is not exclusive. Potter, for instance, uses examples of sexist biases in the practice of science to reveal the ways in which such biases are part of the methods and metaphysics of science.

A connecting theme of Harding, Keller, and Longino's articles is the question, "Are we asking for a feminist science?" According to all three, the answer is a definite "No!" Harding argues against the idea of a distinctive feminist method of research, noting that an important component of many feminist critiques is the rejection of the "methodolatry" of assuming that any one method can encompass the varieties of possible types of knowledge and experience. In place of a quest for a feminist method, Harding offers an examination of the characteristics that account for the fruitfulness and power of contemporary feminist research.

Keller and Longino emphasize the social construction of our category of "feminine" as well as that of "science." Keller notes the parallels between the relation of sex and gender, with that of nature and science. She employs this parallel to illustrate the political construction of our notions of difference, and to call for an alternative understanding that would make viable a notion of "difference in science" rather than a "different science." Longino cautions us against uncritically embracing the notion of a science that values the "feminine," reminding us of the multiplicity of women's experiences, as well as the social construction of our conception of the feminine.

Longino suggests that we shift focus from research programs that are an expression of woman's nature to programs that are consistent with the values and commitments we express in the rest of our lives—from constructing a feminist science to the process of doing science as a feminist. She demonstrates that the idea of a feminist science assumes an absolutism contradictory with the feminist recognition of the ways in which values inform the practice and theory of science.

Irigaray directs her gaze at the language of science, arguing that the sexism of science is so deeply rooted that it is a part of the discourse of science. She illustrates how scientists speak as if without subject, dropping any reference to "I" or "you" or "we." Scientists claim to "let the facts speak for themselves," positing a transcendent position reflecting the desire to be "in front of nature," and thus forgetting that they are *in* nature. While acknowledging that gender biases affect the questions that scientists ask—directing research, say, to female contraception rather than male—Irigaray emphasizes that the very language of science limits the possibilities for discourse. The logic of science allows for such relations as negation, conjunction, disjunction, or implication, but talk of difference, reciprocity, exchange, permeability, or fluidity are hampered or silenced by it. To give voice to these silenced possibilities, Irigaray calls for a language that is differently sexed. We can read Ginzberg's "Uncovering Gynocentric Science" in this context as an attempt to give voice to such a way of speaking of science.

Ginzberg argues that an alternative practice of science, what she calls gynocentric science, has a long history but has been overlooked because it differs in significant ways from traditional science. Suggesting that love and eros are central to the epistemology of this science, Ginzberg reexamines midwifery as a paradigm of gynocentric science.

The essays of Alcoff and Heldke focus upon the epistemological underpinnings of contemporary feminist critiques of science. Alcoff, like Harding and Longino, cautions against the easy assumption that by removing all elements of masculine bias from science we will have purified science. Such a position assumes an objectivist epistemology inconsistent with numerous feminist perspectives. (Irigaray here serves as an excellent example.) Alcoff also critiques the opposite movement into a radical relativism. She offers for discussion two models of theory-choice employed by feminist theorists to undercut both absolutism and radical relativism: the Holistic Model, which she traces through Peirce, Quine, and Hesse, to Harding, Potter, and Keller; and the Constructivist Model of Foucault and Gadamer, exemplified by the work of Westkott, Smith, Hartsock, and de Lauretis.

Heldke continues the dialogue begun by Alcoff by offering a detailed discussion of the epistemology of Evelyn Fox Keller. Showing similarities between Keller's epistemology and that of John Dewey, Heldke argues that Keller's epistemology involves neither absolutism nor relativism. One similarity between the epistemologies of Keller and Dewey illustrated by Heldke is a respect for difference. Through her discussion, Heldke offers a valuable analysis of the concept of difference, a concept that plays a central role in Keller's article.

I begin the section on feminist critiques of the practice of science with

Hubbard's reflections upon the nature of a science consistent with feminist values. Desiring a science in which people take responsibility for the facts that are generated, Hubbard considers the alternatives of a science *for* the people and a science *by* the people. To emphasize the need for such alternatives, she illustrates the variety of ways in which people get excluded from science. From the social structure of the laboratory, to the ideology of woman's nature, to the gender bias of the language of science, Hubbard points out the political content and role of contemporary science. Agreeing with Harding, Longino, and Alcoff that politics is an inherent part of any science, Hubbard calls for a science to which more people have access and a process of validation that is under public scrutiny.

Potter addresses the question, "Is sexist science bad science?" She argues that the question itself presupposes an objectivity and absolutism inconsistent with many feminist perspectives, and blurs our understanding of the gender politics of theory construction. She advocates the adoption of an extended Network Model of theory-construction, or what Alcoff labeled the Holistic Model. Through a detailed examination of the circumstances surrounding the adoption of the corpuscular philosophy in England, Potter illustrates the ways in which theory generation can be affected by gender politics.

The next four essays offer case studies of the effects of gender bias on the practice of science. My essay examines the history of reproductive theory from Aristotle to the seventeenth century. I argue that the adherence to a belief in the inferiority of the female creative principle biased scientific perception of the nature of woman's role in human generation. The essay by the Biology and Gender Study Group continues this discussion by examining the narratives of nineteenth- and twentieth-century science concerning fertilization and sex determination. The Biology and Gender Study Group illustrate that the myths of male activity and female passivity, woman as incomplete, and the male as the true parent inform the models and metaphors of contemporary reproductive theories. They argue that alternative interpretations are made available by rejecting such biases.

Zita examines the evidence used to posit the existence of a premenstrual syndrome and the causal explanations designed to account for it. She traces a pattern of gender biases in the observations and theory construction surrounding PMS research. Zita offers an excellent example of the type of "context stripping" that is the subject of Irigaray's essay. Genova subjects recent theories of hemispheric specialization and lateralization studies to a feminist critique. She demonstrates that the claims of sex difference arising from such studies involve an unacceptable determinism, an ideological bias against women, and a questionable theory of meaning.

As feminists have come to understand the depth of the impact of the gender system in science, we have realized that surface inequities are often grounded by less visible gender biases in the methods and metaphysic of science. This has led us to question the model of science in which science, done properly, is viewed as the epitome of objectivity. Feminists, in company with other theorists, have rejected this image of science. Science is a cultural institution and as such is structured by the political, social, and economic values of the culture within which it is practiced. Although feminists were not the first to reject the traditional image of science, we were the first to carefully explore the myriad ways in which sexist biases affected the nature and practice of science. The articles in this anthology are part of this important and exciting investigation.

Reference

Weininger, Otto. 1906. *Sex and character*. New York: G. P. Putnam's Sons.

Acknowledgments

This anthology was made possible by the efforts and devotion of many people. The special issues of *Hypatia: A Journal of Feminist Philosophy*, from which these articles were compiled, were the brainchild of Peg Simons. She assisted me at every stage of the process of compiling the anthology. I thank her for her suggestions, encouragement, and hard work; she was always there when I had a problem or a request.

The original issues were a product of the work and commitment of many people. Barbara Imber worked with me on every aspect of production. Her unflagging energy and cheerful support kept me going even during the most hectic periods. I am also indebted to many others who generously gave their time and energy reviewing essays or advising me. My thanks to Kathy Addelson, Paula England, Ann Garry, Judith Genova, Harvey Graff, Donna Haraway, Sandra Harding, Hilde Hein, Sarah Hoagland, Noretta Koertge, Diana E. Long, Helen Longino, Sue V. Rosser, Stephanie Shields, Sheryl St. Germain, and Judith Todd. I am also grateful to Robert Corrigan, who assisted me in obtaining funds for the project.

And to all of the contributors, who put so much love into their work and who gave me energy through their enthusiasm, I thank you.

Part One

Overview

Feminist Scholarship in the Sciences: Where Are We Now and When Can We Expect a Theoretical Breakthrough?

SUE V. ROSSER

The work of feminists in science may seem less voluminous and less theoretical than the feminist scholarship in some humanities and social science disciplines. However, the recent burst of scholarship on women and science allows categorization of feminist work into six distinct but related categories: 1) teaching and curriculum transformation in science, 2) history of women in science, 3) current status of women in science, 4) feminist critique of science, 5) feminine science, 6) feminist theory of science. More feminists in science are needed to further explore science and its relationships to women and feminism in order to change traditional science to a feminist science.

The trickle of scholarship on feminism and science begun in the late 1960s and 70s has widened in the 1980s to a continuous stream of books and articles in journals expressing the new scholarship. The work of feminists in science may seem less voluminous and less theoretical than the feminist scholarship in some humanities and social science disciplines. Men continue to dominate science in terms of numbers (National Science Foundation 1986). Feminist philosophers and historians of science (Fee 1981; 1982; Haraway 1978; Hein 1981; and Keller 1982) have described the specific ways in which the very objectivity said to be characteristic of scientific knowledge and the whole dichotomy between subject and object are, in fact, male ways of relating to the world, which specifically exclude women. This "masculinity" of science has contributed to the difficulty for feminists engaging in scientific scholarship. However, the recent increase in amounts and variety of feminist scientific scholarship may be building the foundations for more theoretical work towards a feminist transformation of science.

Hypatia vol. 2, no. 3 (Fall 1987). © by Sue V. Rosser.

The work done by feminists in science might be divided into six distinct but related categories:

1) Teaching and curriculum transformation in science: From the beginning of the current phase of the women's movement, feminists in science have sought means to include more information about women in the science curriculum and methods to attract women to the study of science. Some of the earlier writing in this area focused on course content, appropriate readings, and syllabi for courses on women's health (Beckwith 1981) which included information not covered in standard science and health curricula. Very soon, courses on gender and science, which dealt with how theories and methods of science are biased by gender or support established assumptions about gender (Fausto-Sterling 1982), were described. Writings on several courses about the history of women in science also emerged at about this time (Kien 1984; Alic 1982). Biologists such as Lowry and Woodhull (1983) studied and wrote about factors in college science programs that bring women to major in science. More recently, scientists (Rosser 1986) have focused upon the varieties of women-related courses taught in different parts of the science curriculum and upon integration of information about women and feminist approaches into the traditional introductory level courses in science (Schuster and Van Dyne 1985; Woodhull et al. 1985; Rosser 1985).

2) History of women in science: Although we sometimes labor under the false impression that women have only become scientists in the latter half of the 20th century, early works by Christine de Pizan ([1405] 1982), Giovanni Boccaccio ([1355–59] 1963), and H. J. Mozans ([1913] 1974) recorded past achievements of women in science. Their works underscore the fact that women have always been in science. However, all too frequently the work of women scientists has been credited to others, brushed aside and misunderstood, or classified as non-science. There are several classic examples of the loss of the names of women scientists and the values of their work. Eli Whitney was the employee of Catherine Green, the person who really invented the cotton gin. Since Whitney applied for the patent, he is credited with the invention of this important piece of technology (Haber 1979). Rosalind Franklin's fundamental work on the x-ray crystallography of DNA, which led to the theoretical speculation of the double helical nature of the molecule by Watson and Crick, continues to be brushed aside and undervalued (Watson 1969; Sayre 1975). The groundbreaking work of Ellen Swallow in water, air and food purity, sanitation, and industrial waste disposal which began the science of ecology was reclassified as home economics primarily because the work was done by a woman (Hynes 1984). Ellen Swallow is thus honored as the founder of home economics rather than as the founder of ecology.

The recovery of the names and contributions of the lost women of science has been invaluable research provided by historians of science who were spurred on by the work of feminists in history. Much of the work has followed the male model, focusing on the great or successful women in science. Olga Opfell's (1978) *The Lady Laureates: Women Who Have Won the Nobel Prize* and Lynn Osen's (1974) *Women in Mathematics* are based upon this model. Many individual biographies on famous figures such as Marie Curie (Reid 1974), Rosalind Franklin (Sayre 1975), Sophie Germain (Bucciarelli and Dworsky 1980), Mary Somerville (Patterson 1983), and Sofia Kovalenskia (Koblitz 1983) have also emerged. Demonstrating that women have been successful in traditional science is important in that it documents the fact that despite the extreme barriers and obstacles, women can do science. This work is what Lerner (1975) calls compensatory history. In the scheme proposed by McIntosh (1983) for transformation of the standard liberal arts curriculum by the new scholarship on women, the works on these exceptional women scientists would be categorized under Phase II: Heroines, exceptional women or an elite few who are seen to have been of benefit to culture as defined by the traditional standards of the discipline.

Some historians have rejected this male model and sought to examine the lives and situations of women in science who were not famous. Margaret Rossiter's (1982) *Women Scientists in America: Struggles and Strategies to 1940* is the groundbreaking work that examines how the work of the usual woman scientist suffers from underrecognition due to application of double standards and other social barriers inherent in the structure of the scientific community. Vivian Gornick (1983) adopts some of this approach in her popularized work *Women in Science* in which she explores the daily lives, hopes and frustrations of contemporary women scientists. From these studies, qualitative information about the status of women in science emerges.

3) Current status of women in science: Quantitative statistical data serve to complete the picture of the situation for women in science. Under the aegis of the Commission on Professionals in Science and Technology (formerly the Scientific Manpower Commission) Betty Vetter and Eleanor Babco collect this data for the American Association for the Advancement of Science on a continuing basis. The National Science Foundation used this same data base for its 1986 comprehensive report *Women and Minorities in Science and Engineering.* The NSf report documents that despite women's increasing participation at every degree level in every field and employment sector, women still face lower salaries at all levels, lower rates of tenure and promotion, and higher rates of unemployment in all fields. Surveys such as those conducted by the National Research Council (biennially every year since 1973) and various professional societies (Women Chemists' Committee of the American Chemical Society 1983) have reinforced the data collected by Vetter and Babco regarding women's poor status in the scientific community.

Two recently edited volumes, *Women in Scientific and Engineering Professions* (Haas and Perrucci 1984) and *Women in Science* (Kahle 1985), feature women scientists discussing different aspects of the problems of the status of women in science. Kahle examines comparative data from Western, Eastern European, and developing countries and considers teaching methods in the secondary schools. She concludes that the best way to attract more women into science is to provide role models, practice with experimental equipment, and information and encouragement about careers for women in science at the elementary and secondary school level (Kahle 1985). Although Harding (1986) suggests that this affirmative action approach should be in some ways quite unthreatening to traditional science, it is to be hoped that the increased numbers of women in science will permit more flexibility in the way in which women pursue science and will provide the opportunity for the development of a feminist science (Rosser 1986).

4) Feminist critique of science: The inferior status of women within the society of science is a reflection of women's inferior status in society at large. Very frequently the data from scientific experiments and theories gleaned from those data are used to provide a scientific basis to justify women's inferior position in society. Feminists of the 19th Century such as Blackwell ([1875] 1976) critiqued Darwin's theory of sexual selection which purported to demonstrate the innate passivity and subordination of women. Zetkin ([1896] 1976] attacked the flaws in the craniometry research which attempted to demonstrate links between intelligence and the research findings that women's brains were smaller and lighter (Broca 1873), or had more or less prominent areas (Patrick 1895) than those of men. Tanner (1896), Thompson (1903), and Calkins (1896) pointed out the biased assumptions in the early research on intelligence which demonstrated women's inferiority in this area also.

Recent feminists have drawn attention to the experimental biases, flaws in experimental design, and overinterpretation of data which occur in the 20th century areas of research. Janet Sayer's (1982) *Biological Politics* explores the links between the areas of research begun in the 19th century and their 20th century counterparts. Bleier (1979), Lowe (1978), Rosser (1982), Birke (1986) and other feminists in science have refuted the biologically deterministic theory known as Sociobiology which contends that genes determine behavior, therefore providing a biological basis for women's position in society. They have discussed the problems inherent in extrapolating from one species to another in behavioral traits and the circularity of logic involved with using human language and frameworks to interpret animal behavior which is then used to "prove" that certain human behavior is biologically determined since it has also been found in animals.

Birke discusses the ways in which biologically deterministic arguments have been used to limit women and justify our socially inferior position in society. She notes that acceptance of biological determinism may lead to conservative political policies that justify existing social arrangements. In extreme cases biological determinism lends itself to biological/medical interventions to change behaviors such as homosexuality and violence. For example, the deterministic explanation implies that techniques such as hormone therapy and surgery may change lesbians and male homosexuals to heterosexuals.

Bleier (1979), Star (1979) and Sayers (1982) have critiqued the studies on brain lateralization, hormones and brain anatomy (Bleier 1984) that attempt to link differences in male and female brains with behavioral traits such as visuo-spatial ability, verbal ability and aggression. These feminists have demonstrated flaws in experimental design, assumptions based on limited experimental data, unwarranted extrapolation of data from rodents to humans, and problems with biochemical conversion of hormones from estrogens to androgens within the body (Bleier 1984). Fennema and Sherman (1977), Sherman (1980) and Haven (1972) have documented the importance of social factors such as number of mathematics courses taken, familiarity with games that develop visuospatial skills and peer pressure as extremely important factors affecting the differential performance of adolescent males and females on tests of mathematical ability. Feminists (Hubbard and Lowe 1979) as well as other scientists (Lewontin, Rose and Kamin 1984) have pointed out the cultural and gender biases in the I. Q. test.

In several recent works, *Science and Gender* (Bleier 1984), *Reflections on Gender and Science* (Keller 1985), *Women, Feminism and Biology* (Birke 1986) and *Myths of Gender* (Fausto-Sterling 1985), feminists who are scientists critique the research in all of these areas. They conclude that faulty research design, overextension of data and general "bad" science may be more acceptable in these areas of research because the areas provide scientific bases for the social status quo. This research tends to generate scientific data to support the superior status of the white western male over white women and all people of color. Most of the work of feminists in science has understandably been focused on demonstrating the unscientific nature of these biologically deterministic theories.

In her recent work, *The Science Question in Feminism*, the philosopher of science Sandra Harding (1986) discusses five effects that the feminist critique has had on science. "First of all, equity studies have documented the massive historical resistance to women's getting the education, credentials, and jobs available to similarly talented men; they have also identified the psychological and social mechanisms through which discrimination is informally maintained even when the formal barriers have been eliminated" (Harding 1986, 21). Second, the critique has revealed the use of science to support "sexist, racist, homophobic, and classist social projects." Third, the critiques question the extent to which all science must be value-laden and biased towards men's perspective both in selection and definition of research problems and in the design and interpretation of research. Fourth, the feminist critique has used techniques from other disciplines such as psychoanalysis, literary criticism and historical interpretation to reveal "the hidden symbolic and structural agendas—of purportedly value-neutral claims and practices" (Harding 1986, 23) of science. Finally, feminist epistemologies provide an alternative understanding for what kinds of social experience "should ground the beliefs we honor as knowledge" (Harding 1986, 24).

After discussing the evidence for each of these effects Harding concludes that the feminist critiques that point out the androcentrism, and therefore "bad science," raise a paradox. "Clearly, more scientifically rigorous and objective inquiry has produced the evidence supporting specific charges of androcentrism but that same inquiry suggests that this kind of rigor and objectivity is androcentric" (Harding 1986, 110). This paradox in turn raises the question of the potential for a non-androcentric or even gynocentric science.

5) Feminine science: Although the bulk of work of feminists in science has been in the category of a critique of science, recently feminists have begun to raise the question of whether or not women do science differently from men. Many feminists have argued that gender is socially constructed and that there are no biologically-based differences between men and women in traits, behaviors or abilities. Social construction of gender differences implies that those differences might be obliterated or emphasized based upon changes in social policies and programs. In contrast, feminists such as Elshtain (1981) and MacMillan (1982) reject social construction of gender for biological construction of gender. Biological determinism as espoused by these feminists proposes that women have gender-based traits that are to be valued and that grant a superiority to women. But Birke sees biological determinism as fraught with the same problems whether proposed by feminists or antifeminists. She takes a middle-of-the-road feminist approach; on this issue she accepts the significance of biology, but not biological determinism, in interaction with social factors to determine gender. Considering the information from the previous three categories, even when biological determinism is put aside, it is clear that the history and status of women in science is different from that of men. That differential history and status, as well as differences in the socialization of men and women in our society, may provide women with a perspective outside that of the dominant culture (i.e., a feminist perspective) from which to critique science.

It is also quite possible that these differences may have allowed at least two other factors to operate which may make the work that women scientists do different from the work done by men scientists. First, since science is considered a masculine pursuit in our culture (for a discussion of science as a masculine province see Keller 1982; Hein 1981; Fee 1982; Harding 1986), science being done by women is often defined as non-science. Hynes (1984) documents the redefinition of Ellen Swallow's experiments in water chemistry, toxicity and food purity out of the science of chemistry and into home economics. Her work was considered unscientific because it was interdisciplinary research done by a woman (Hynes 1984). In *For Her Own Good*, Ehrenreich and English (1979) explore the ways in which methods to aid birthing are considered to be science or non-science depending upon who is practicing them. Midwifery is not usually considered a science because it is practiced by women although obstetrics is a science because it is a medical field dominated by men. Thus, in order to fully appreciate the contributions of women to science, the definition of science may need to be broadened or questioned so that some areas defined as women's fields are considered.

Second, some recent work by feminists studying women scientists may indicate that the approaches taken by women scientists and the theories generated by those approaches may be qualitatively different from those of male scientists. In her study of the life and work of Barbara McClintock, *A Feeling for the Organism*, E. F. Keller (1983) suggests that McClintock adopted a more feminine or less objective approach towards the object of her study. In her approach studying maize, McClintock shortens the distance between the observer and the object being studied and considers the complex interaction between the organism and its environment. Her statement upon receiving the Nobel Prize was that "it might seem unfair to reward a person for having so much pleasure over the years, asking the maize plant to solve specific problems and then watching its responses" (Keller 1983). This statement suggests a closer, more intimate relationship with the subject of her research than typically is expressed by the male "objective" scientist. One does not normally associate words such as "a feeling for the organism" with the rational, masculine approach to science. McClintock also did not accept the predominant hierarchical theory of genetic DNA as the "Master Molecule" that controls gene action, but focused on the interaction between the organism and its environment as the locus of control. Keller (1983) suggests that this interactive, non-hierarchical model may represent a theory that is more reflective of feminine values than that posed by the Watson-Crick model of DNA as the "Master Molecule."

6) Feminist theory of science: In describing McClintock's approach as more feminine than that taken by most of the male practitioners of science, Keller raises the question of whether gender can influence the methods and theories of science. This is a question which feminists in science are now beginning to debate. McClintock herself does not believe that gender (including her own) affects the methods and theories of science. Even Keller has stated that good science is gender neutral.

> My exploration of McClintock's life has also sharpened my thinking on a subject I have written about elsewhere: the relation of gender to science.

> In her adamant rejection of female stereotypes, McClintock poses a challenge to any simple notions of a "feminine" science. Her pursuit of a life in which "the matter of gender drops away" provides us instead with a glimpse of what a "gender-free" science might look like. (Keller 1983, xvii).

Feminists in disciplines in the humanities and social sciences have developed feminist theories that have transformed the theoretical frameworks upon which their discipline is based. In anthropology, for example, feminists have re-evaluated the role of woman the gatherer (Dahlberg 1981) in relationship to man the hunter, questioned concepts such as dominance and subordination as the only effective ways to describe complex social relationships among people (Leacock 1977), and re-examined the relationship between gender and the evolution of tool use. Feminist anthropologists brought about the awareness that other approaches and alternative approaches that may even run counter to the established assumptions can yield more information and provide a more complete picture of reality.

Feminist historians and philosophers of science (Hein 1981; Haraway 1978; and Fee 1982) have suggested that a theory for a feminist science might also develop. Fee has stated that a sexist society should be expected to develop a sexist science. Conceptualizing a feminist science from within our society is "like asking a medieval peasant to imagine the theory of genetics or the production of a space capsule" (Fee 1982, 31).

Some feminists who are scientists have even sketched some of the parameters that might distinguish a feminist science from the traditional science as it is now practiced in our culture. Bleier (1984) suggests that central ideas to a feminist science might be the rejection of dualisms such as subjectivity/objectivity, rational/feeling and nature/culture which focus our thinking about the world. Kahle (1985) believes that changes in elementary and secondary school teaching which will attract more women to science might lead to a feminist science. Fee (1986) has hinted that feminist visions of science are similar to visions proposed by other groups who differ in race, class, or tradition rather than gender from the white middle or upper class Western men who are the major developers of scientific theories. Fee's comment again raises the question of whether or not gender has a particular influence on science. Posed in broader terms, the question is whether or not "good" science really can be objective, or whether, since it is a human pursuit, it will always be laden with human values and biases such as those of race, class and gender. Birke's vision is perhaps more far-reaching in that she considers the interactive relationship between feminism and science. She recognizes that science will not be changed by feminism as long as feminists remain solely in the passive, critical role. We must be active and engaged in science to cause the change to occur. "If science is to be changed, then it has to be done in ways that take account of gender and other differences, and that does mean that feminists have to engage actively in making that change" (Birke 1986, 171).

The demand of Birke (1986) is for feminists to work with science and attempt to change it. She recognizes the ultimate failure of feminist theories and feminists who reject science and technology because they have been used to dominate and control women. Rejecting science and technology will not stop its influence in our daily lives and its capabilites for death and destruction. The only way that feminists can really hope to change it is to understand science and its relationships to women and feminism.

Perhaps what is needed to further explore the relationship between science and gender is more of the type of exploration suggested by the word play implied by the title of Harding's (1986) work: *The Science Question in Feminism*. Feminists from a variety of disciplines must examine the effect that feminism has had on their discipline; simultaneously many feminists from an individual discipline looking at the particular methods and insights that might be brought to the study of feminism from that discipline may begin to further untangle the complicated interrelationships between feminism and science, as well as other disciplines. What we need are more feminist scientists, feminist historians, feminist philosophers and feminists in every discipline.

References

American Chemical Society. 1983. Medalists study charts women chemists' role. *Chemistry and Engineering* (Nov. 15): 53.

Alic, M. 1982. The history of women in science: A women's studies course. *Women's Studies International Forum* 3 (1): 75–81.

Beckwith, B. 1981. Women empowering women. *Science for the People* 13 (5): 19–22.

Birke, L. 1986. *Women, feminism and biology. The feminist challenge*. New York: Methuen.

Blackwell, A. B. [1875] 1976. *The sexes throughout nature*. New York: G. P. Putnam's Sons. Reprint. Westport, CT: Hyperion Press.

Bleier, R. 1979. Social and political bias in science: An examination of animal studies and their generalizations to human behavior and evolution. In *Genes and Gender II*, eds. R. Hubbard and M. Lowe. Staten Island, NY: Gordian Press.

———. 1984. *Science and gender: A critique of biology and its theories on women*. New York: Pergamon Press.

Boccaccio, G. [1355–59] 1963. De Claris Mulieribus. In *Concerning famous women*, trans. Geudo A. Guarino. New Brunswick, New Jersey.

Broca, P. 1873. Reported in *Nature* 8: 152.

Bucciarelli, L. and N. Dworsky. 1980. *Sophie Germain: An essay in the history of the theory of elasticity*. Dordrecht, Holland: D. Reidel.

Calkins, M. W. 1896. Community of ideas of men and women. *Psychological Review* 3 (4): 426–430.

Dahlberg, F. 1981. *Woman the gatherer*. New Haven: Yale University Press.

de Pizan, C. [1405] 1982. *The book of the city of ladies*. Reprint, slightly modified from the translation by Earl Jeffrey Richards. New York: Persea Books.

Ehrenreich, B. and D. English. 1979. *For her own good: 150 years of the experts advice to women*. New York: Doubleday.

Elshtain, J. B. 1981. *Public man, private woman: Woman in social and political thought*. Princeton, NJ: Princeton University Press.

Fausto-Sterling, A. 1982. Teaching aids: Focus on women and science. Course closeup: The biology of gender. *Women's Studies Quarterly* 10 (2): 17–19.

———. 1985. *Myths of gender: Biological theories about men and women*. New York: Basic Books.

Fee, E. 1981. Is feminism a threat to scientific objectivity? *International Journal of Women's Studies* 4 (4): 213–233.

———. 1982. A feminist critique of scientific objectivity. *Science for the People* 14 (4): 8.

———. 1986. Critiques of modern science: The relationship of feminism to other radical epistemologies. In *Feminist approaches to science*, ed. Ruth Bleier. New York: Pergamon Press.

Fennema, E. and J. Sherman. 1977. Sex-related differences in mathematics achievement, spatial visualization and affective factors. *American Educational Research Journal* 14: 51–71.

Gornick, V. 1983. *Women in science: Portraits from a world in transition*. New York: Simon and Schuster.

Haas, V. and C. Perruci. 1984. *Women in scientific and engineering professions*. Ann Arbor: University of Michigan Press.

Haber, L. 1979. *Women pioneers of science*. New York: Harcourt Brace Jovanovich.

Haraway, D. 1978. Animal sociology and a natural economy of the body politic, Part I: A political physiology of dominance; and Animal sociology and a natural economy of the body politic, Part II: The past is the contested zone: Human nature and theories of production and reproduction in primate behavior studies. *Signs: Journal of Women in Culture and Society* 4 (1): 21–60.

Harding, S. 1986. *The science question in feminism*. Ithaca, NY: Cornell University Press.

Haven, E. W. 1972. Factors associated with the selection of advanced mathematical courses by girls in high school. *Research Bulletin*, 72, 12. Princeton: Educational Testing Service.

Hein, H. 1981. Women and science: Fitting men to think about nature. *International Journal of Women's Studies* 4 (4): 213–233.

Hubbard, R. and M. Lowe. 1979. Introduction. In *Genes and gender II*, eds. R. Hubbard and M. Lowe. New York: Gordian Press.

Hynes, H. P. 1984. Women working: A field report. *Technology Review* (Nov./Dec.): 3Bff.

Kahle, J. B. 1985. *Women in science: A report from the field*. Philadelphia: Taylor and Francis.

Keller, E. 1982. Feminism and science. *Signs: Journal of Women in Culture and Society* 7 (3): 589–602.

———. 1983. *A feeling for the organism: The life and work of Barbara McClintock*. New York: W. H. Freeman and Company.

———. 1985. *Reflections on gender and science*. New Haven: Yale University Press.

Kien, J. and D. Cassidy. 1984. The history of women in science: A seminar at the University of Regensburg, FRG. *Women's Studies International Forum* 7 (4): 313–317.

Koblitz, A. H. 1983. *A convergence of lives, Sofia Kovlevskia: Scientist, writer, revolutionary*. Cambridge, MA: Birkhauser Boston.

Leacock, E. 1977. Women in egalitarian societies. In *Becoming visible: Women in European history*, eds. R. Bridenthal and C. Koonz. Boston: Houghton Mifflin.

Lerner, G. 1975. Placing women in history: A 1975 perspective. *Feminist Studies* 3 (1–2): 5–15.

Lewontin, R. C., S. Rose, and L. Kamin. 1984. *Not in our genes: Biology, ideology, and human nature*. New York: Pantheon Books.

Lowe, M. 1978. Sociobiology and sex differences. *Signs: Journal of Women in Culture and Society* 4 (1): 118–125.

Lowry, N. and A. Woodhull. 1983. New directions in science education. *Science for the People* 15 (1): 31–36.

MacMillan, C. 1982. *Woman, reason and nature*. Princeton, NJ: Princeton University Press.

McIntosh, P. 1983. Interactive phases of curricular re-vision: A feminist perspective. Working paper No. 124, Wellesley College, Center for Research on Women, Wellesley, MA.

Mozans, H. J. 1974. *Women in science-1913*. Cambridge, MA: MIT Press.

National Science Foundation. 1986. *Women and minorities in science and engineering*, Report 86–300.

Opfell, O. 1978. *The lady laureates: Women who have won the Nobel prize*. New Jersey: Methuen.

Osen, L. M. 1974. *Women in mathematics*. Cambridge, MA: MIT Press.

Patrick, G. T. W. 1895. The Psychology of Women. *Popular Science Monthly* 47: 209–25.

Patterson, E. 1983. *Mary Somerville and the cultivation of science 1815–1840*. The Hague: Nijhoff.

Reid, R. 1974. *Marie Curie*. New York: Mentor Books.

Rosser, S. V. 1982. Androgyny and sociobiology. *International Journal of Women's Studies* 5 (5): 435–444.

———. 1985. Introductory biology: Approaches to feminist transformations in course content and teaching practice. *Journal of Thought, An Interdisciplinary Quarterly* 20 (3): 205–217.

———. 1986. *Teaching science and health from a feminist perspective: A practical guide*. New York: Pergamon Press.

Rossiter, M. W. 1982. *Women scientists in America: Struggles and strategies to 1940*. Baltimore: The Johns Hopkins University Press.

Sayers, J. 1982. *Biological politics*. London: Tavistock.

Sayre, A. 1975. *Rosalind Franklin and DNA: A vivid view of what it is like to be a gifted woman in an especially male profession*. New York: W. W. Norton.

Schuster, M. and S. Van Dyne. 1985. Women's place in the academy: *Transforming the liberal arts curriculum*. Rowman and Allanheld.

Sherman, J. 1980. Mathematics, spatial visualization, and related factors: Changes in girls and boys, grades 8–11. *Journal of Educational Psychology* 72: 476–482.

Star, S. L. 1979. The politics of right and left: Sex differences in hemispheric brain asymmetry. In *Women look at Biology looking at women*, eds. R. Hubbard, M. S. Henifin and B. Fried. Boston: Schenkman.

Tanner, A. 1896. The community of ideas of men and women. *Psychological Review* 3 (5): 548–550.

Thompson (Woolley), H. B. 1903. *The mental traits of sex*. Chicago: University of Chicago Press.

Watson, J. D. 1969. *The double helix*. New York: Atheneum Publishers, Mentor Paperback.

Woodhull, A. M., N. Lowry, and M. S. Henifin. 1985. Teaching for change: Feminism and the sciences. *Journal of Thought: An Interdisciplinary Quarterly* 20 (3): 162–173.

Zetkin, C. [1896] 1976. Only with the proletarian woman will socialism be victorious. Reprinted in *The Socialist Register*. H. Draper and A. G. Lipow, eds. London: Merlin Press.

Part Two

Feminist Theories of Science

Is There a Feminist Method?

SANDRA HARDING

A continuing concern of many feminists and non-feminists alike has been to identify a distinctive feminist method of inquiry. This essay argues that this method question is misguided and should be abandoned. In doing so it takes up the distinctions between and relationships among methods, methodologies and epistemologies; proposes that the concern to identify sources of the power of feminist analyses motivates the method question; and suggests how to pursue this project.

Over the last two decades the moral and political insights of the women's movement have inspired social scientists and biologists to raise critical questions about the ways traditional researchers have explained gender, sex, and relations within and between the social and natural worlds. From the beginning, criticisms of traditional research methods have generated speculations about alternative feminist methods. Is there a feminist method of research that can be used as a criterion to judge the adequacy of research designs, procedures and results? Is there a distinctive method that should be taught to students wanting to engage in feminist inquiry? If so, what is it?

My point in this paper is to argue against the idea of a distinctive feminist method of research. I do so on the grounds that preoccupation with method mystifies what have been the most interesting aspects of feminist research processes. Moreover, I think that it is really a different concern that motivates and is expressed through most formulations of the method question: what is it that makes some of the most influential feminist-inspired biological and social science research of recent years so powerful? I begin by indicating some variants of the method question and their diverse origins.[1]

ORIGINS OF QUESTIONS ABOUT METHODS

We can now see that many androcentric assumptions and claims have survived through the years within, but unquestioned by most researchers in, traditional areas of biology and the social sciences.[2] In pondering the sources

Hypatia vol. 2, no. 3 (Fall 1987).

of such partial and distorted accounts, feminists have criticized the favored methods of their disciplines. They argue that prescription of only a narrow range of methods has made it impossible to understand women's "natures" and lives or how gender activities and commitments influence behaviors and beliefs. For instance, it is a problem that most social scientists have not questioned the practice—indeed, have sometimes overtly prescribed it—of having only men interview only men about both men's and women's beliefs and behaviors. It is a problem that biologists prescribe that cosmetics be tested only on male rats on the grounds that the estrus cycle unduly complicates experiments on female rats. It is a problem that Lawrence Kohlberg did not wonder about why he was finding women's responses to his moral dilemmas less easy than the men's responses to sort into the categories he had set up. As Carol Gilligan pointed out, Kohlberg thereby replicated the assumption of Freud, Piaget and Ericson that the patterns of moral development characteristic of men should be regarded as the model of human moral development (Gilligan 1982). Faulty methods of inquiry appear to be implicated in these and many other such cases.

Moreover, it is not just particular research methods that are the target of feminist criticisms, but the fetishization of method itself. The practice in every field of choosing research projects because they can be completed with favored methods amounts to a kind of methodolatry (in Mary Daly's memorable phrase) that sacrifices scientific explanation and increased understanding to scientistic fashions in research design (Daly 1973, 11). For example, Carolyn Wood Sherif's survey of changing fashions in methods during the last half-century of psychology documents how this field attempted to gain legitimacy through blind imitation of what it wrongly took to be scientific methods. In imitating only the style of natural science research, psychology has produced distorted images of its subject matters (Sherif 1979). Historians of science join Sherif in arguing that the attempt to legitimate a new field in inquiry by claiming that it uses scientific method even more rigorously than its predecessors frequently masks scientisms of the flimsiest sort (e.g., Fee 1980).

Responding to these criticisms of mainstream methods, a number of proposals for alternative feminist methods have emerged: I mention three that are well-known. Catharine MacKinnon (1982) has argued that "consciousness-raising *is* feminist method." (She does so in the course of important criticisms of attempts to fit feminist analyses of rape and other acts of violence against women within the constraints of liberal and marxist legal theory.) Even though this claim captures an intuitively important truth about feminist thought, it leaves biologists and social scientists puzzled about what, exactly, they should *do* differently in their laboratories, offices or field work. MacKinnon is thinking of method in ways very different from uses of the term in "methods" courses in the social sciences or in lab work in biology. While this is certainly not a good reason to reject or ignore MacKinnon's point, it does motivate a closer examination of just what role unique methods play in feminist inquiry.

Nancy Hartsock (1983) argues for a "specifically feminist historical materialism." By replicating in the title of this paper components of the method of Marxist political economy ("dialectical historical materialism"), she appears to hold that it is a method of inquiry as well as an epistemology she is explicating. Other social scientists have also pointed to the benefits of adapting Marxist method to the needs of feminist inquiry (Hartmann 1981). Are methodology and epistemology also at issue in such discussions of method? It would be reasonable to argue that these analyses provide methodologies for feminism, not methods for feminist inquiry. That is, they show how to adapt the general structure of Marxist theory to a kind of subject matter Marxist political economy never could fully appreciate or conceptualize. Is it method, methodology, theory or epistemology that is the issue here? Is there any point to trying to distinguish these in such cases?

A third candidate for the feminist method arises from sociologists who favor phenomenological approaches. They appear to argue that feminist method is whatever is the opposite of excessive empiricism or of positivist strains in social research. In particular they focus on the virtues of qualitative vs. quantitative studies, and on the importance of the researcher identifying with, rather than objectifying, her (women) subjects (e.g., Stanley and Wise 1983). These critics draw attention to misogynistic practices and scientistic fetishes in social inquiry that frequently have had horrifying consequences for both theory and public policy. However, the prescription of a phenomenological approach as *the* feminist method ignores many of the well-known problems with such ways of conducting social research. (These are problems also for approaches to social inquiry labelled intentionalist, humanistic or hermeneutical.) For example, there are things we want to know about large social processes—how institutions come into existence, change over time and eventually die out—that are not visible through the lens of the consciousnesses of those historical actors whose beliefs and activities constitute such processes. Moreover, the subjects of social inquiry are often not aware even of the local forces that tend to shape their beliefs and behaviors, as Freud, for instance, pointed out. The explanation of "irrational" beliefs and behaviors is just part of this subject matter that is not amenable to phenomenological inquiry processes alone. Furthermore, these approaches do not fully enough appreciate the importance of critical studies of masculinity and men (in the military, the economy, the family, etc.); nor of feminist critical studies of women's—including feminists'—beliefs and behaviors. Finally, what would it mean for biology to use these approaches? Or, is feminism to insist anew on just the division between nature and culture it has been so successful in problematizing?

A different source of debate over method arises from the fact that men and women social scientists who want to begin to contribute to the feminist research agendas sometimes express puzzlement about whether there is a special method which must be learned in order to do feminist research. If so, what is it? Their disciplinary literatures are peppered with references to "empirical method," "phenomenological method," and sometimes even "marxist method," so it seems reasonable to seek the feminist method. Such a method obviously should be included as one of the alternative "research methods" taught in every social science.

No doubt some of these researchers are also perplexed about how to conceptualize alternatives to the very common, though (they are willing to grant) sexist practices of their disciplines because they have been taught to think about methods of inquiry as the general categories of concrete research practices. As one sociologist protested, there are only three methods of social research: listening to what people say (in response to questions or in spontaneous speech), observing what they do (in laboratory situations or in the field, collectively or individually), and historical research. Therefore, she argued, there can't be a feminist method of inquiry. And yet she would admit that the problem of eliminating androcentric results of research appears to require more than simply having well-intentioned individuals conduct research, since we know that individual intent is never sufficient to maximize objective inquiry. After all, where is the feminist inquirer who has not come to understand the inadequacy of some of her or his own earlier practices and beliefs? What are the implications of the feminist critiques and counter proposals for conceptions of method such as this common one?

Finally, it is the use of a particular method of inquiry—scientific method—that is supposed to make the results of Galileo's and Newton's research so much more valuable than those of the medieval theologians whose accounts of astronomy and physics modern science dislodged. If feminist researchers are criticizing scientific method, it appears to some that they must be proposing some *non-scientific* method of inquiry. While this train of thought usually emerges from critics hostile to feminism, at least traces of it appear in the thinking of some feminists. The latter are outraged—and properly so—at the distortions and subsequent policy crimes committed in the name of science. But they cannot see how to reform or transform science so as to eliminate those consequences that are so damaging to women. Accepting scientists' frequent claim that the true foundations of science are to be found in its distinctive method, they claim that science and its method are part of the problem against which feminism must struggle.

These discussions, meditations, claims and analyses create a rich and confusing field for thinking about method. I turn next to some traditional distinctions between methods, methodologies and epistemologies—and to some traditional confusions about some of these.

METHOD VS. METHODOLOGY VS. EPISTEMOLOGY

Philosophers have favored a distinction between method and methodology.[3] Will this distinction help us to sort out what is at issue in the method question?

Method. A research method is a technique for gathering evidence. Thus, the feminist sociologist's focus on listening to what subjects say, observing them or searching through historical records names categories of concrete techniques for gathering evidence. How unique to feminism are the techniques of evidence gathering used in the most widely acclaimed examples of feminist research? Is it to these techniques that we can attribute the great fecundity and power of feminist research in biology and the social sciences? Carol Gilligan listens to what women say when presented with familiar examples of moral dilemmas—a traditional method of inquiry, according to the sociologist cited earlier. But she uses what she hears to construct a powerful challenge to the models of moral development favored by Western psychologists (and philosophers) (Gilligan 1982). Sociologists observe men's and women's uses of public spaces—a traditional way of gathering evidence. But they conclude that suburbs, usually thought of as desirable worlds for women, are in fact hostile to women's needs and desires (Werkerle 1980). Joan Kelly-Gadol examines ancient property and marriage records—yet another traditional research method—and goes on to argue that women did not have a renaissance, at least not during the Renaissance. The progress of (some) men appears to require loss of status for all women at those moments Western cultures evaluate as progressive (Kelly-Gadol 1976). Obviously some of the most influential feminist researchers have used some very traditional methods. Of course they do begin their research projects by listening more skeptically to what men say and more sympathetically to what women say; they observe both men and women with new critical awarenesses; they ask different questions of history. We shall explore further how best to characterize these kinds of differences between feminist and traditional research. But at this point it is reasonable to conclude that in a way of characterizing research methods that is common in the social sciences, it is traditional methods that these feminist researchers have used.

This line of argument at least has the virtue of making feminist research appear acceptable as social *science* to those who had doubts. After all, it is using familiar methods of research: what more could one ask? However, this benefit is sometimes inadvertently bought at the cost of providing support for continued willful ignorance about feminist research on the part of such skeptics. If there is a way of arguing that there is "nothing new here," such critics can say to themselves and to others: why read it, teach it, be concerned with it at all? As a political strategy, there are both benefits and risks to the way of thinking about feminist research methods that I have been proposing.

Androcentric critics are not the only audience for answers to the method question; and they will, no doubt, rarely run out of self-convincing reasons to continue in their ways. Feminists, too, have claimed to identify a distinctive method in feminist research processes, and it would be a service to our own discourses if we could gain greater clarity about whether we should be making such claims. Can we conclude at this point that there is both less and more going on in feminist inquiry than new methods of research? The "less" is that it is hard to see how to characterize what is new about these ways of gathering evidence in terms of methods. The "more" is that it is new methodologies, epistemologies and other kinds of theories that are requiring new research processes.

Methodology. Traditional philosophers have contrasted methods with methodologies. A *methodology* is a theory and analysis of "the special ways in which the general structure of theory finds its application in particular scientific disciplines" (Caws 1967, 339). Here is surely one major place we should look to discover what is distinctive about feminist research. At least since Kuhn (1970) it has been clear that methods of inquiry can not be regarded as independent of the general theories, specific hypotheses and other background assumptions that guide research (optical assumptions guiding inferences from observations; logical assumptions about the appropriate structure of inferences; assumptions about computers, questionnaires and other scientific "instruments"; assumptions about the veracity of historical archives; etc.) (Harding 1976). Surely it is the great innovations in feminist theory that have led to gathering evidence in different ways—even if this evidence gathering still falls into such broad categories as listening to what people say, observing them and historical inquiry. For example, discussions of how sociobiology, or psychoanalytic theory, or Marxist political economy, or phenomenology should be used to understand various aspects of women and gender addressed by different disciplines are methodology discussions that lead to distinctive research agendas—including ways of gathering evidence. Feminist researchers have argued that these theories have been applied in ways that make it difficult to understand women's participation in social life, or to understand men's activities as gendered (vs. as representing "the human"). They have transformed these theories to make them useful in such neglected projects.

The traditional definition of methodology with which I began here makes it appear that all theoretical discussions should be subsumed under the heading "methodology." But isn't a discussion about how to apply marxist categories to analyses of women's role in reproduction different from a later discussion that assumes the results of the earlier one and goes on to hypothesize what that role is? Clearly the choice of a substantive feminist theory of any kind will have consequences for what can be perceived as appropriate research methods. Do excessively empiricist (positivist) anti-theoretical tendencies in the social sciences make some feminist researchers here find it more comfortable to talk about competing "methodologies" rather than about competing theories? Should the fundamental distinctions here be between methods, substantive theories (and "methodology" discussions will help select these) and epistemologies? Not a great deal turns on answers to these questions—I certainly do not wish to replicate traditional philosophical fetishization of distinctions. But asking such questions heightens awareness of the extent to which our thinking is still constrained by distorting remnants of positivism.

Epistemology. It might be assumed that for a philosophical audience little needs to be said about what an epistemology is. It is, of course, a theory of knowledge, "concerned with the nature and scope of knowledge, its presuppositions and basis, and the general reliability of claims to knowledge" (Hamlyn 1967, 8–9). But the fact that everyone could agree to this definition masks some important conflicts between feminist epistemological activity and traditional ways of thinking about the subject. For one thing, the mainstream Anglo-American tradition has sharply distinguished epistemology from philosophy of science. The former is construed as analyses of the nature and scope of "ordinary" knowledge. The latter is construed as analyses of the nature and scope of explanation for the natural sciences in general and/or for particular sciences. Insistence on this rigid distinction shields both fields from the kinds of epistemological questions arising from feminist scientific research. It makes it difficult to recognize those questions as either epistemological or as ones in the philosophy of science since they openly link the two domains. (Of course, it is also a problem for many traditional thinkers that these questions arise from social science and biology, not from physics or astronomy, and that they are feminist.) Such thinkers forget how the contours of modern epistemology have been designed and redesigned in response to the thought of Copernicus, Galileo, Newton and later natural and social scientists.

Feminist epistemologies are responses to at least three problems arising from research in biology and the social sciences. In the first place, if scientific method is such a powerful way of eliminating social biases from the results of research, as its defenders argue, how come it has left undetected so much sexist and androcentric bias? Is scientific method impotent to detect the presence of certain kinds of widespread social bias in the processes and results of inquiry? In the second place, is the idea of woman as knower a contradiction in terms? The issue here is not the existence of individual women physicists, chemists, astronomers, engineers, biologists, sociologists, economists, psychologists, historians, anthropologists, etc. There have been many of these throughout the history of science, and they have made important contributions to this history (Alic 1986; Rossiter 1982). Instead, the issue is that knowledge is supposed to be based on experience; but male-dominance has simultaneously insured that women's experience will be different from men's, and that it will not count as fruitful grounds from which to generate scientific problematics or evidence against which to test scientific hypotheses. Feminist research in biology and the social sciences has "discovered" and made good use of this lost grounds for knowledge claims. But the idea that more objective results of research can be gained by grounding research in distinctively gendered experience is a problem for traditional thinking in epistemology and the philosophy of science—to understate the issue. Finally, the more objective, less false, etc. results of feminist research clearly have been produced by research processes guided by the politics of the women's movement. How can the infusion of politics into scientific inquiry be improving the empirical quality of the results of this research? ("Think of Lysenkoism! Think of Nazi science!")

These kinds of questions about knowers and knowledge appear to be ruled out of court by this popular Anglo-American philosophical definition of what is to count as epistemology and as philosophy of science. In fact these are fundamental epistemological questions in the sense of the encyclopedia definition above, and they are ones crucial to the understanding of what has counted as adequate grounds for belief in the history of scientific explanation.

There is a second reason to pause before assuming that traditional philosophy has an adequate grasp of the nature of epistemology. Sociologists of knowledge have pointed to the benefits to be gained from a kind of "externalist" understanding of epistemologies that raise skeptical questions about the adequacy of philosophers' "internalist" understandings—to borrow a contrast from the history of science. The sociologists argue that epistemologies are best understood as historically situated strategies for justifying belief, and that the dominant ones in mainstream philosophy of science fail to consider important questions about the "scientific causes" of generally accepted beliefs, including those in the history of science (Bloor 1977; Latour 1987; Latour and Woolgar 1979). They criticize the assumption that only false beliefs have social causes, and they call for accounts of the social conditions generating historically situated true beliefs as well as false ones. These sociologies of knowledge have all the flaws of functionalist accounts; they implicitly assume as grounds for their own account precisely the epistemology they so effectively undermine; and they are bereft of any hint of feminist consciousness.[4] Nevertheless, one can recognize the value for feminist theorists of an approach that opens the way to locating within history the origins of true belief as well as false belief, and also epistemological agendas within science. Justificatory strategies have authors, and intended and actual audiences. How do assumptions about the gender of audiences shape traditional epistemological agendas and claims? How should feminists reread epistemologies "against the grain?" How should such sociological and textual understandings of epistemologies shape feminist theories of knowledge?

This is perhaps the best place to point out that there have been at least three main epistemologies developed in response to problems, such as those identified earlier, arising from the feminist social science and biological research. Each draws on the resources of an androcentric tradition and attempts to transform the paternal discourse into one useful for feminist ends. What I have called "feminist empiricism" seeks to account for androcentric bias, women as knowers, and more objective but politically guided inquiry while retaining as much as possible of the traditional epistemology of science. What their authors call the "feminist standpoint epistemologies" draw on the theoretically richer resources of marxist epistemology to explain these three anomalies in ways that avoid the problems encountered by feminist empiricism. But postmodernist critics argue that this epistemology founders on the wrecks of no longer viable enlightenment projects. Feminist forms of postmodernism attempt to resolve these problems, in the process revealing flaws in their own paternal discourse. I have argued that the flaws in these epistemologies originate in what they borrow from their non-feminist ancestors, and that in spite of these flaws, each is effective and valuable in the justificatory domains for which it was intended (Harding 1986).

Epistemological concerns have not only arisen from feminist research in biology and the social sciences but also generated it. So we have traced what is distinctive in feminist research back from techniques of evidence gathering, to methodologies and theories, and now to new hypotheses about the potency of scientific method, about the scientific importance of women's experience, and about the positive role some kinds of politics can play in science.

One final comment is in order here. It is a problem that social scientists tend to think about methodological and epistemological issues primarily in terms of methods of inquiry—for example, in "methods courses" in psychology, sociology, etc. A-theoretical tendencies and excessively empiricist tendencies in these fields support this practice. No doubt it is this habit that tempts many social scientists to seek a unique method of inquiry as the explanation for what is unusual about feminist analyses. On the other hand, it is also a problem that philosophers use such terms as "scientific method" and "the method of science" when they are referring to issues of methodology and epistemology. Unfortunately, it is all too recently in the philosophy of science that leading theoreticians meant by the term some combination of deduction and induction—such as making bold conjectures that are subsequently to be subjected to severe attempts at refutation. The term is vague in yet other ways. For instance, it is entirely unclear how one could define "scientific method" so that it referred to practices common to every discipline counted as scientific. What methods do astronomy and the psychology of perception share? It is also unclear how one would define this term in such a way that highly trained but junior members of research teams in physics counted as scientists, but farmers in simple societies (or mothers!) did not (Harding 1986, ch. 2).

What is really at issue in the feminist method question? No doubt more than one concern. But I think that an important one is the project of identifying what accounts for the fruitfulness and power of so much feminist research. Let us turn to address that issue directly.

WHAT DO FEMINIST RESEARCHERS DO?

Let us ask about the history of feminist inquiry the kind of question Thomas Kuhn posed about the history of science (Kuhn 1970, Ch. 1). He asked what the point would be of a philosophy of science for which the history of science failed to provide supporting evidence. We can ask what the point would be of elaborating a theory of the distinctive nature of feminist inquiry that excluded the best feminist biological and social science research from satisfying its criteria. Some of the proposals for a feminist method have this unfortunate consequence. Formulating this question directs one to attempt to identify the characteristics that distinguish the most illuminating examples of feminist research.

I shall suggest three such features. By no means do I intend for this list to be exhaustive. We are able to recognize these features only after examples of them have been produced and found fruitful. As research continues, we will surely identify additional characteristics that expand our understandings of what makes feminist research explanatorily so powerful. No doubt we will also revise our understandings of the significance of the three to which I draw attention. My point is not to provide a definitive *answer* to the title question of this section, but to show that this historical approach is the best strategy if we wish to account for the distinctive power of feminist research. While these features have consequences for the selection of research methods, there is no good reason to call them methods.

The "Discovery" of Gender and its Consequences. It is tempting to think that what is novel in feminist research is the study of women, or perhaps even of females. It is certainly true that women's "natures" and habits frequently have been ignored in biological and social science research, and that much recent inquiry compensates for this lack. However, since Engels, Darwin, Freud and a vast array of other researchers have in fact studied women and females—albeit partially and imperfectly—this cannot in itself be a distinguishing feature of feminist inquiry.[5]

However, it is indeed novel to analyze gender. The idea of a systematic social construction of masculinity and femininity that is little, if at all, constrained by biology, is very recent. One might even claim that contemporary feminism has "discovered" gender in the sense that we can now see it everywhere, infusing daily beliefs and behaviors that were heretofore thought to be gender-neutral. In board rooms and bedrooms, urban architecture and suburban developments, virtually everywhere and anytime we can observe, the powerful presence and vast consequences of gender now appear in plain sight. Moreover, feminist accounts examine gender critically. They ask how gender—and, especially, tensions between its individual, structural and symbolic expressions—accounts for women's oppression (Harding 1986, Ch. 2). They also ask how gendered beliefs provide lenses through which researchers in biology and social science have seen the world. That is, gender appears as a variable in history, other cultures, the economy, etc. But, like class and race, it is also an analytic category within scientific and popular thought, and through which biological and social patterns are understood. So, feminist research is distinctive in its focus on gender as a variable and an analytic category, and in its critical stance toward gender.

Does this special subject matter of feminist inquiry have consequences for choices of methods of inquiry? Traditional philosophies of science often try to separate ontological assumptions about the appropriate domain of inquiry from issues about methods of inquiry. And yet, one can note that behaviorists' insistence that the proper subject matter of psychology is human "matter in motion"—not the mental states hiding out in the "black boxes" of our minds—leads them to develop distinctive research designs elaborating their notions of experimental method. Certainly the shifts in subject matters characteristic of feminist research have implications for method selection even if one cannot understand what is so powerful in feminist research by turning first to its methods.

Women's Experiences as a Scientific Resource. Reflection on this "discovery" of gender leads to the observation that gender—masculinity as well as femininity—has become a problem requiring explanation primarily from the perspective of women's experiences. A second distinctive feature of feminist research is that it generates its problematics from the perspective of women's experiences. It also uses these experiences as a significant indicator of the "reality" against which hypotheses are tested. It has designed research *for* women that is intended to provide explanations of social and biological phenomena that women want and need (Smith 1979). The questions that men have wanted answered all too often have arisen from desires to pacify, control, exploit or manipulate women, and to glorify forms of masculinity by understanding women as different from, less than, or a deviant form of men. Frequently, mainstream social science and biology have provided welfare departments, manufacturers, advertisers, psychiatrists, the medical establishment and the judicial system with answers to questions that puzzle these men in these institutions, rather than to the questions women need answered in order to participate more fully, and have their interests addressed, in policy decisions. As the sociologist Dorothy Smith (1979) points out, all too often social science works up everyday life into the conceptual forms necessary for administrative and managerial forms of ruling: its questions are administrative/managerial ones.

In contrast, feminist inquiry asks questions that originate in women's experiences. Angela Davis (1971) asks why Black women's important role in the family, and the Black family's in Black culture, are either invisible or distorted in the thinking of white (male) policy makers, Black male theorists and white feminist theorists.[6] Carol Gilligan (1982) asks why male theorists have been unable to see women's moral reasoning as valuable. Joan Kelly-Gadol (1976) asks if women really did share in the "human progress" taken to characterize the Renaissance. Ruth Hubbard (1983) asks if it is really true, as mainstream interpretations of evolutionary theory would lead one to believe, that only men have evolved. Women's perspective on our own experiences provide important empirical and theoretical resources for feminist research. Within various different feminist theoretical frameworks, they generate research problems and the hypotheses and concepts that guide research. They also serve as a resource for the design of research projects, the collection and interpretation of data, and the construction of evidence.[7]

Is the selection of problematics part of scientific method? On the one hand, traditional scientists and philosophers will prohibit the inclusion of these parts of the processes of science "inside science proper." But that is only because they have insisted on a rigid distinction between contexts of discovery that are outside the domain of control by scientific method and contexts of justification where method is supposed to rule. This distinction is no longer useful for delimiting the area where we should set in motion objectivity-producing procedures, since it is clear that we need to develop strategies for eliminating the distortions in the results of inquiry that originate in the partial and often perverse problematics emerging from the context of discovery. Restriction of "science proper" to what happens in the contexts of justification supports only mystifying understandings of how science has proceeded in history. In reaction to this tendency, it is tempting to assert that these features of science are part of its method.

On the other hand, I have been arguing that there is no good reason to appropriate under the label of method every important feature of the scientific process. How hypotheses arrive at the starting gate of "science proper" to be tested empirically turns out to make a difference to the results of inquiry. Therefore, this process must be critically examined along with everything else contributing to the selection of evidence for or against the maps of the world that the sciences provide. We have identified here another distinctive feature of feminist inquiry that has consequences for the selection of ways to gather evidence, but that is not itself usefully thought of as a research method.

A Robust Gender-sensitive Reflexivity Practice. A third feature contributing to the power of feminist research is the emerging practice of insisting that the researcher be placed in the same critical plane as the overt subject matter, thereby recovering for scrutiny in the results of research the entire research process. That is, the class, race, culture and gender assumptions, beliefs and behaviors of the researcher her/himself must be placed within the frame of the picture that she/he paints. This does not mean that the first half of a research report should engage in soul searching (though a little soul searching by researchers now and then can't be all bad!). Indeed, a more robust conception of reflexivity is required than this. To come to understand the historical construction of race, class and culture within which one's subject matter moves requires reflection on the similar tendencies shaping the researcher's beliefs and behaviors. Understanding how racism shaped opportunities and beliefs for Black women in 19th century United States requires reflection on how racism shapes opportunities and beliefs for white social scientists and biologists in 1987. Moreover, feminist social scientists (and philosophers!) bear gender too; so a commitment to the identification of these historical lenses in our own work—not just in the work of 19th century theorists or of men—will also help avoid the false universalizing that has characterized some feminist work. Thus the researcher appears in these analyses not as an invisible, anonymous, disembodied voice of authority, but as a real, historical individual with concrete, specific desires and interests—and ones that are sometimes in tension with each other. Since these characteristics of the researcher are part of the evidence readers need to evaluate the results of research, they should be presented with those results (See Smith 1979). We are only beginning to be able to elaborate the rich dimensions of feminist subjectivities, but this project promises to undermine Enlightenment assumptions that distort traditional theory and much of our own thinking (Martin and Mohanty 1986).

Like the other features that I have proposed as distinguishing marks of the best of feminist inquiry to date, this robust reflexivity is not helpfully contained within categories of method.

Conclusion

I ended an earlier section of this essay at a point where our search for influences on the selection of research methods had led us to methodologies and substantive theories, and then to epistemologies. After considering what some of the most influential feminist research has done, we are in a position to see how deeply politics and morals are implicated in the selection of epistemologies, and consequently of theories, methodologies and methods. Many of the most powerful examples of feminist research direct us to gaze critically at all gender (not just at femininity), to take women's experiences as an important new generator of scientific problematics and evidence, and to swing around the powerful lenses of scientific inquiry so that they enable us to peer at our own complex subjectivities as well as at what we observe. To follow these directions is to violate some of the fundamental moral and political principles of Western cultures. Thus meditation on the method question in feminism leads us to the recognition that feminism is fundamentally a moral and political movement for the emancipation of women. We can see now that this constitutes not a problem for the social science and biology that is directed by this morals and politics, but its greatest strength. Further exploration of why this is so would lead us to a more elaborate discussion of the new understandings of objectivity feminism has produced—a topic I cannot pursue here (see Harding 1986).

It may be unsatisfying for many feminists to come to the conclusion that I have been urging: the search for a distinctive feminist method of inquiry is not a fruitful one. As a consolation, I suggest we recognize that if there were some simple recipe we could follow and prescribe in order to produce powerful research and research agendas, no one would have to go through the difficult and sometimes painful—if always exciting—processes of learning how to see and create ourselves and the world in the radically new forms demanded by our feminist theories and practices.

NOTES

1. Parts of this essay appear in an earlier form as the introduction to Harding 1987.
2. Many of the points I shall make are applicable to some areas of biology as well as to the social sciences. I shall mention biology in these cases.
3. See Caws' (1967) discussion of the history of the idea of scientific method and of this distinction.
4. These claims would have to be argued, of course. It is not difficult to do so. Rose (1979), and Rose and Rose (1979) have made one good start on such arguments.
5. I mention this because some traditional researchers erroneously think they are teaching a feminist course or doing feminist inquiry simply by "adding women" or females to their syllabi, samples or footnotes.
6. Davis is just one of the theorists who has led us to understand that in cultures stratified by race and class, there is no such thing as just plain gender, but always only white vs. Black femininity; working class vs. professional class masculinity, etc.
7. As I have discussed elsewhere, they do not provide the final word as to the adequacy of these various parts of research processes. Nor is it only women who have produced important contributions to feminist understandings of women's experiences (Harding 1986, 1987).

REFERENCES

Alic, Margaret. 1986. *Hypatia's heritage: A history of women in science from antiquity to the late nineteenth century*. London: The Women's Press.

Bloor, David. 1977. *Knowledge and social imagery*. London: Routledge & Kegan Paul.

Caws, Peter. 1967. Scientific Method. In *The encyclopedia of philosophy*, Volume 7, ed. P. Edwards. New York: Macmillan Publishing Co.

Daly, Mary. 1973. *Beyond god the father: Toward a philosophy of women's liberation*. Boston: Beacon Press.

Davis, Angela. 1971. Reflections on the black woman's role in the community of slaves. *Black Scholar* 3: 2–15.

Fee, Elizabeth. 1980. Nineteenth century craniology: The study of the female skull. *Bulletin of the History of Medicine* 53.

Gilligan, Carol. 1982. *In a different voice: Psychological theory and women's development*. Cambridge, MA: Harvard University Press.

Hamlyn, D. W. 1967. History of epistemology. In *The encyclopedia of philosophy*, Volume 3, ed. P. Edwards. New York: Macmillan Publishing Co.

Harding, Sandra, ed. 1976. *Can theories be refuted? Essays on the Duhem-Quine thesis*. Dordrecht: D. Reidel.

———. 1986. *The science question in feminism*. Ithaca, NY: Cornell University Press.

———, ed. 1987. *Feminism and methodology: Social science issues*. Bloomington: Indiana University Press.

Harding, Sandra and Merrill Hintikka, eds. 1983. *Discovering reality: Feminist perspectives on epistemology, metaphysics, methodology and philosophy of science*. Dordrecht: D. Reidel.

Harding, Sandra and Jean O'Barr, eds. 1987. *Sex and scientific inquiry*. Chicago: University of Chicago Press.

Hartmann, Heidi. 1981. The unhappy marriage of marxism and feminism. In *Women and revolution*, ed. L. Sargent. Boston: South End Press.

Hartsock, Nancy. 1983. The feminist standpoint: Developing the ground for a specifically feminist historical materialism. In *Discovering reality*. See Harding and Hintikka 1983.

Hubbard, Ruth. 1983. Have only men evolved? In *Discovering reality*. See Harding and Hintikka 1983.

Kelly-Gadol, Joan. 1976. Did women have a renaissance? In *Becoming visible: Women in European history*, eds. R. Bridenthal and C. Koonz. Boston: Houghton Mifflin.

Kuhn, Thomas S. 1970. *The structure of scientific revolutions*. 2nd ed. Chicago: University of Chicago Press.

Latour, Bruno. 1987. *Science in action*. Milton Keynes, England: Open University Press.

Latour, Bruno and Steve Woolgar. 1979. *Laboratory life: The social construction of scientific facts*. Beverly Hills, CA: Sage.

MacKinnon, Catharine. 1982. Feminism, marxism, method and the state:

An agenda for theory. *Signs: Journal of Women in Culture and Society* 7 (3): 515–544.

Martin, Biddy and Chandra Talpade Mohanty. 1986. Feminlst politics: What's home got to do with it? In *Feminist studies/Critical studies*, ed. Teresa de Lauretis. Bloomington, IN: Indiana University Press.

Rose, Hilary. 1979. Hyper-reflexivity: A new danger for the countermovement. In *Countermovements in the sciences*, eds. Helga Nowotny and Hilary Rose. Dordrecht: D. Reidel.

Rose, Hilary and Steven Rose. 1979. Radical science and its enemies. In *Socialist register*, eds. Ralph Miliband and John Saville. Atlantic Highlands, NJ: Humanities Press.

Rossiter, Margaret. 1982. *Women scientists in America: Struggles and strategies to 1940*. Baltimore: Johns Hopkins University Press.

Sherif, Carolyn. 1979. Bias in psychology. In *The prism of sex: Essays in the sociology of knowledge*, eds. Julia A. Sherman and Evelyn Torton Beck. Madison: University of Wisconsin Press.

Smith, Dorothy. 1979. A sociology for women. In *The prism of sex*. See Sherif 1979.

Stanley, Liz and Sue Wise. 1983. *Breaking out: Feminist consciousness and feminist research*. Boston: Routledge & Kegan Paul.

Werkerle, Gerda R. 1980. Women in the urban environment. *Signs: Journal of Women in Culture and Society* 5:3 Supplement.

The Gender/Science System: or, Is Sex to Gender as Nature Is to Science?

EVELYN FOX KELLER

In this paper, I explore the problematic relation between sex and gender in parallel with the equally problematic relation between nature and science. I also offer a provisional analysis of the political dynamics that work to polarize both kinds of discourse, focusing especially on their intersection (i.e., on discussions of gender and science), and on that group most directly affected by all of the above considerations (i.e., women scientists).

The most critical problem facing feminist studies today is that of the meaning of gender, its relation to biological sex on the one hand, and its place with respect to other social markers of difference (e.g., race, class, ethnicity, etc.) on the other—i.e., the relation between sex, gender and difference in general. Similarly, I would argue that the most critical problem facing science studies today is that of the meaning of science, its relation to nature, and its place with respect to other social institutions—i.e., the relation between nature, science, and interests in general. My purpose in this paper is, first, to identify some important parallels, even a structural homology, between these two questions, and second, to suggest that an exploration of this homology (including the factors responsible for its maintenance) can provide us with some useful guidelines in our attempts to address these problems.

Three different kinds of parallels between feminist studies and science studies can be identified immediately: one might be called historical, the second, epistemological, and the third, political. Historically, it is worth noting that modern feminist studies actually emerges with the recognition that women, at least, are made rather than born—i.e, with the distinction between sex and gender. In much the same way, contemporary studies of science come into being with the recognition of a distinction between science

Hypatia vol. 2, no. 3 (Fall 1987). © by Evelyn Fox Keller.

and nature—with the realization that science not only is not now, but can never be, a "mirror of nature." With the introduction of these distinctions came the growth of two new (essentially non-overlapping) fields of study, one devoted to the analysis of the social construction of gender, and the other, of the social construction of science.

In both of these endeavors, however, scholars now find they must contend with what might best be described as a "dynamic instability"[1] in the basic categories of their respective subjects: in the one case, of gender, and the other, of science—an instability in fact unleashed by the very distinctions that had given them birth. If gender is not to be defined by sex, nor science by nature (i.e., by what *is*), how then *are* they to be defined? In the absence of an adequate answer to this question, the difficulties that both feminist and science scholars have encountered in maintaining yet containing their necessary distinctions (between sex and gender on the one hand, and between science and nature on the other) are as familiar as they have been insurmountable. It is this phenomenon that I am calling the epistemological parallel. Just as discussions of gender tend to lean towards one of two poles—either toward biological determinism, or toward infinite plasticity, a kind of genderic anarchy, so too do discussions of science exhibit the same polarizing pressures—propelled either towards objectivism, or towards relativism. In one direction, both gender and science return to a premodern (and pre-feminist) conception in which gender has been collapsed back onto sex, and science, back onto nature. Under the other, we are invited into a post-modernist, post-feminist (and post-scientific) utopia in which gender and science run free, no longer grounded either by sex or by nature—indeed, in which both sex and nature have effectively disappeared altogether. Attempts to occupy a "middle ground"—either with respect to gender or to science—must contend not only with the conceptual difficulty of formulating such a position, but also with the peculiarly insistent pressures of a public forum urging each concept toward one pole or the other.

I invoke the label "political" for the third parallel with reference to the politics of knowledge: I believe that, finally, what is at issue in both cases is a question of status—the status, in the one case, of gender, and in the other case, of science, as theoretical categories. Is gender, as an analytical category, different from, perhaps even prior to, categories of race, class, etc.? (Or, to reverse the question, is oppression ultimately the only important variable of gender?) The parallel question is, of course: is science substantively different from other social structures or "interest groups?"; i.e., are scientific claims to knowledge any better than other (non-scientific) claims to knowledge?

The parallels I am describing between feminist studies and social studies of science have in fact been just that, i.e., parallels; until quite recently, there has been virtually no intersection between the two disciplines, just as there has been virtually no interaction between attempts to reconceptualize gender and science—as if the two categories were independent, each having nothing to do with the other. It is only with the emergence of a modern feminist critique of science that the categories of gender and science have come to be seen as intertwined, and, accordingly, that the two subjects (feminist studies and science studies) have begun to converge. But, as I have argued elsewhere, this most recent development actually required the prior conceptualization both of gender as distinct from sex, and of science as distinct from nature. That is, the modern feminist critique of science is historically dependent on the earlier emergence of each of its parent disciplines. With such a lineage, however, it also (perhaps necessarily) inherits whatever ambiguity/unclarity/uncertainty/instability remains in each of the terms, gender and science. Indeed, it might be said that feminist studies of science has become the field in which these ambiguities are most clearly visible, and accordingly, the field that offers the best opportunity for understanding the factors that may be working against a clear and stable "middle ground" account of both concepts. I suggest also that, for this reason, an examination of the history of feminism and science can provide important insights to help point the way towards resolution of these difficulties.

To illustrate these claims, I will focus on one particular episode in the recent history of feminism and science, namely on contemporary debate over the idea of a "feminist science," and more specifically, over the question of whether or not the recently celebrated cytogeneticist, Barbara McClintock, might be regarded as an exemplar of such a "feminist science." In order to orient this debate, however, I need to preface the discussion with a few very brief remarks about the prior history of the struggles of that group most directly affected by the issues of feminism and science, namely women scientists.

Throughout this century, the principal strategy employed by women seeking entrance to the world of science has been premised on the repudiation of gender as a significant variable for scientific productivity. The reasons for this strategy are clear enough: experience had demonstrated all too fully that any acknowledgement of gender based difference was almost invariably employed as a justification for exclusion. Either it was used to exclude them from science, or to brand them as "not-women"—in practice, usually both at the same time. For women scientists *as scientists*, the principal point is that measures of scientific performance admitted of only a single scale, according to which, to be different was to be lesser. Under such circumstances, the hope of equity, indeed, the very concept of equity, appeared—as it still appears—to depend on the disavowal of difference. In hindsight, it is easy enough to see the problem with this strategy: If a universal standard invites the translation of difference into inequality, threatening further to collapse into duality, otherness, and exclusion, the same standard invites the translation of equality into sameness, and accordingly, guarantees the exclusion of any experience, perception or value that is other. As a consequence, "others" are eligible for inclusion only to the extent that they can excise those differences, eradicating even the marks of that excision. Unfortunately, such operations are often only partially successful, leaving in their wake residual handicaps that detract from the ability of the survivors to be fully effective "competitors." Yet more importantly, they fail to provide effective protection against whatever de facto discrimination continues to prevail. What such a strategy *can* do however is help obscure the fact of that discrimination.

This dilemma is perhaps nowhere more poignantly illustrated than in the experiences of women scientists in the mid 20th century—the nadir of the history of women in American science. Having sought safety in the progressive eradication of any distinguishing characteristics that might mark their gender, by the 1950's, women scientists, qua women, had effectively disappeared from American science. Their numerical representation was no longer recorded; even, by their own choice, their tell-tale first names were withheld from publications. Unfortunately, however, this strategy failed to protect actual women from the effects of an increasingly exclusionary professional policy—it only helped obscure the effects of that policy.

The principal point here is that these women were caught on the horns of an impossible dilemma—a dilemma that was unresolvable as long as the goal of science was seen as the unequivocal mirroring of nature, and its success as admitting of only a single standard of measurement. It was only with the introduction of an alternative view of science—one admitting of a multiplicity of goals and standards—that the conditions arose for some feminists, in the late 1970's and early 80's, to begin to argue for the inclusion of difference—in experience, perceptions, and values—as intrinsically valuable to the production of science. Very rapidly, however, the idea of difference in science gave way (in some circles) to the extremely problematic idea of a different science altogether—in particular, to the idea of a feminist science. A feminist science might mean many things to different people, but, in practice, it is almost always used to invoke the idea of a "feminine" science. As it actually happened, though, for the idea of such a feminist/feminine (or "femininist") science to really engage people's fancy, something more than simply the availability of an alternative to the traditional view of science was needed: a source of legitimation was required, and even better, an exemplar. In this need, the Nobel Prize Committee seemed fortuitously to oblige.

In 1983, that committee selected Barbara McClintock for its award in medicine and physiology for work done almost forty years earlier. With help from an enthusiastic press, they thereby turned a deviant and reclusive cytogeneticist—a woman who has made respect for difference the cornerstone of her own distinctive philosophy of science—into a new, albeit reluctant, cultural heroine. And with a small but crucial rewriting of the text, advocates of a new "femininist" science found in that moment what they needed: an exemplar who, through her "feeling for the organism," seemed to restore feminine values to science, and who (even more importantly), after years of struggle, had finally been validated, even vindicated, by a reluctant establishment.

Curiously enough, however, at the very same moment, this same establishment was busy welcoming McClintock back into their fold. Now it was their turn to repudiate the differences that in the past had made her such an anomaly. As Stephen Jay Gould informed us (New York Review, Mar. 26, 1984), McClintock's "feeling for the organism" is in no way distinctive; all good scientists (himself included) have it and use it. Other reviewers went further. They had always appreciated her work; the claim that she had been misunderstood, unappreciated, was simply wrong. At the very moment that (some) feminists claimed McClintock as one of their own, as a representative of a different science, mainstream scientists closed ranks around her, claiming her as one of their own: there is only one science.

Because I cannot claim to have been an innocent bystander in these developments, I need to say something about my own contribution. In writing my book on McClintock, it seemed important to me to consciously bracket the questions on gender and science that I had been writing about before, and to treat the material on McClintock's life and work, and their relation to the history of modern biology, in their own terms. It seemed unfair to McClintock's own story to burden it with a prior moral of mine, particularly in view of the fact that it was extremely unclear to me just what the relevance of a feminist, or gender, critique to that story might be. McClintock herself is not a feminist; she has throughout her life not only resisted but adamantly repudiated all classification, and her commitment to science as a place where "the matter of gender drops away" is staunch. If gender was to prove an important variable in that story, particularly to an understanding of her scientific deviance, the story itself would have to show both me and the reader how. This stance of course left readers free to draw their own conclusions, which indeed they did—though, perhaps predictably, usually attributing those conclusions to me.

Partly in self-defense, and partly because it needed to be addressed, I turned to the question of a feminist science in my subsequent book (Keller 1985)—focusing in particular on the question of the relevance of gender to the McClintock story. On quite general grounds, I argued that if one means by a feminist science a feminine science, the very notion is deeply problematic: first because it ignores the fundamentally social character of the process by which both science and scientists get named as such, and second because of the extent to which our understanding of "feminine" and "scientific" have been historically constructed in opposition to each other.

Other problems arise in relation to the McClintock story itself: first, there is not only McClintock's own disavowal of all stereotypic notions of femininity, but, in addition, there is the fact that none of the dynamics we think of as key to feminine socialization seem to apply to her. She was never pregnant, never parented, and, although she was a daughter, her relation to her mother was so anomalous as to pose a challenge to conventional assumptions about mother-daughter bonding. Finally, a directly biological account of McClintock's difference won't do because she is in fact not representative of women in general—not even of women scientists; nor are her vision and practice of science absent among male scientists. What then are we to make of the fact that so much of what is distinctive about that vision and practice—its emphasis on intuition, feeling, connectedness, and relatedness—conform so well to our most familiar stereotypes of women? And are, in fact, so rare among male scientists?

To answer this question, I argued that it was necessary to shift the focus first from sex to gender, and second, from the construction of gender to the construction of science. The question then becomes, not why McClintock relies on intuition, feeling, a sense of connectedness and relatedness in her scientific practice, but how come these resources are repudiated by stereotypic science? Put this way, the question virtually answers itself: the repudiation of these resources, I argued, derives precisely from the conventional naming of science as masculine, coupled with the equally conventional naming of these resources as feminine. The relevance of gender in the McClintock story thus shifts from its role in her personal socialization to its role in the social construction of science. For the project of reclaiming science as a gender-free endeavor—a place where "the matter of gender drops away"—I did however suggest one respect in which McClintock's sex may in fact have provided her with an advantage: "However atypical she is as a woman, what she is *not* is a man"—and hence is under no obligation to prove her "masculinity" (Keller 1985, 174).

In other words, I attempted to articulate the very kind of "middle ground" stance with regard to gender that I have in this paper suggested is so peculiarly difficult (for all of us) to maintain—claiming in particular that the relevance of gender to science is (a) a socially constructed relevance, but (b) *carried* by the sex of its participants (in the sense, that is, that gender specific norms are internalized along with one's "core gender identity"). In short, I argued that gender is a fundamentally relational construct which, although not determined by sex, is never entirely independent of it. In spite of cultural variability and psychological plasticity, it means *something*—though, for many individuals, perhaps not a great deal—to identify oneself as being of one sex and not of another.

Because the responses this argument has generated provide ready evidence of the instability to which our understanding of gender is generally subject, it may be useful to review them here. Many feminists have continued to read the McClintock story as a manifesto of a "feminist science" (in the sense, i.e., of a specifically female science)—in the process, either celebrating me as its proponent, or, if they respond at all to my disclaimers, implying that I lack the courage of my convictions—sometimes even suggesting that I lack the courage of McClintock's convictions. On the other side, a number of readers (both men and women), complained that I ignored McClintock's distinctiveness in attempting to make her "an exemplar of women," or that my argument has to be wrong because of the fact that there *are* men who do think like McClintock, who do have "a feeling for the organism." Hence, they conclude, gender cannot be relevant to this story. Finally, there are those readers for whom the anxiety raised by any reference to gender whatever in the context of science is so great that they simultaneously read both "gender" and "gender-free" as "female" (see, e.g., Koblitz N.d.). For these last readers, especially, the suggestion of even a tenuous link between gender and sex for either women or men is taken as proof of their worst suspicions—i.e., that this argument constitutes a threat to the claim of women scientists to equity.

What is perhaps most notable about these readings is the extent to which they all depend, albeit in different ways, on an unwitting, almost reflexive, equation between questions about gender with questions about sex. That is, they assume that what is really at issue is not the force of gender ideology, but the force of sex. With the space between sex and gender thus eliminated, the original question of McClintock's difference automatically reduces to the exceedingly problematic (and dubious) question of whether or not men and women, by virtue of being male or female, think differently. Not surprisingly, the responses to this last question are both mixed and extremely fervent. Once read as duality, the question of difference becomes subject to only two responses: yes or no, i.e., either the embrace or the denial of duality—embrace by those who welcome it; denial by those who fear it. The entire spectrum of difference has thus been collapsed onto two poles—duality and universality.

I suggest that the conceptual collapse illustrated here—the difficulty so many of us have in thinking about difference in any other terms than either duality or universality—is rooted not in biology but in politics: not a consequence of any limitations in the way in which our brains are constructed, but rather the consequence of an implicit contest for power. I am suggesting that duality and universality are responses actually structured by, as well as employed in, a contest that is first and foremost political. As I've already implied, the Nobel Prize plays a crucial role in creating out of McClintock's science a zone of contention in the first place, inviting both feminists and mainstream scientists to claim her as one of their own. The legitimation and authority provided by the Nobel Prize endowed McClintock's scientific practice with a value worth fighting over—a value claimed by one side by the negation of difference, and by the other by its reformulation as duality. It is precisely in the context of such a competition that the question of difference itself becomes a contested zone—our conceptualization of difference molded by our perceptions (as well as the reality) of power. In other words, in this context at least, the very debate between duality and universality both presupposes and augments a prior division between an "us" and a "them," bound in conflict by a common perception of power. It refers not to a world in which we and they could be said to occupy truly separate spheres, with separate, noninteracting sources of power and authority—in such a world, there would be no debate—but, rather, to a world perceived as ordered by a single source (or axis) of power that is at least in principle commonly available; a world in which duality can be invoked (by either side) to create not so much a separation of spheres as an inside and an outside—in other words, as a strategy of exclusion. It is in just such a world that the perennially available possibility of difference becomes a matter to contest—invoked or denied according, first, to the value attached to that difference, and second, to one's position relative to the axis of power. With nothing to lose, but possibly something to gain, a difference of value can safely and perhaps usefully be claimed as a mark of duality. Standing inside the circle, however, it is more strategic to assimilate any difference that is known to be valuable, and to exclude, through the invocation of duality, those differences that promise no value.

In the particular case at hand, the power at stake is, to put it quite simply, the epistemic authority of scientists. And although I have so far been speaking as if there were only one demarcation capable of effecting exclusion, namely that between men and women, it is neither the case that all women are without scientific authority nor that all those without scientific authority are women. Another demarcation is also operative here—indeed, I would even say, primary: namely, the demarcation between science and non-science, potentially at least as exclusionary as that between men and women. It is the threat of this second demarcation that polarizes our discussions about the nature of scientific knowledge in much the same way as the first polarizes our discussions of gender. The question of whether scientific knowledge is objective or relative is at least in part a question about the claim of scientists to absolute authority. If there is only one truth, and scientists are privy to it, (i.e., science and nature are one), then the authority of science is unassailable. But if truth is relative, if science is divorced from nature and married instead to culture (or "interests"), then the privileged status of that authority is fatally undermined. With this move, the demarcation between science and non-science appears to have effectively dissolved. Because the notion of a feminist/feminine science engages both these demarcations simultaneously—indeed, it could be sald that the very proposal of a feminist science depends on the possibility of playing one off against the other—an understanding of this debate requires that we pay attention to both sets of dynamics. Once again, an examination of the responses of women in science is especially instructive, for it is that group of individuals that is most directly positioned by these two demarcations—indeed, positioned by their intersection.

For women who have managed to obtain a foothold within the world of science, the situation is particularly fraught. Because they are "inside," they have everything to lose by a demarcation along the lines of sex that has historically only worked to exclude them. And precisely because they are rarely quite fully inside, more commonly somewhere near the edge, the threat of such exclusion looms particularly ominously. At the same time, as scientists, they have a vested interest in defending a traditional view of science—perhaps, because of the relative insecurity of their status, even more fiercely than their relatively more secure male colleagues. On two counts then, it is hardly surprising therefore that most women scientists (as well as historians and philosophers of science) vehemently resist the notion of a feminist/feminine science: the suggestion that women, as a class, will do a different kind of science simultaneously invokes the duality of sex, and undermines (or presupposes the undermining of) our confidence in the privileged attachment of science to nature.

What *is* surprising is the extent to which so many women scientists have been able to read into McClintock's Nobel Prize the possibility of an alternative to the classical dichotomies. The McClintock story is compelling to many women working in science because it testifies for them the viability of difference within the world of science as we know it. They read the Nobel Prize not as an invitation to rebellion, but as evidence of the legitimacy of difference within the established criteria of scientific "truth"—as making room within the prevailing canon for many of the questions, methodologies and interpretations that their more familiar version of that canon did not permit. In this, they seek a larger canon rather than a different one; a richer, perhaps even multi-faceted, representation of reality, but not a separate reality. Their resistance to the reduction of difference to duality is firm, and it is, admittedly, a resistance clearly in the service of their own interests as women scientists. But it is also in the service of a larger interest, and that is the preservation of some meaning to the term "science."

Even accepting that the scientific endeavor is not as monolithic as the received view would have it, accepting that science does not and cannot "mirror" nature, the question remains, what do we mean by "science?" Does, indeed, any meaning of the term remain? If it does, that meaning must derive from the shared commitment of scientists to the pursuit of a maximally reliable (even if not faithful) representation of nature, under the equally shared assumption that, however elusive, there is only one nature. We may now realize that science is not capable of apprehending reality "as it is," as people once thought it could, but belief in the existence of separate realities (a notion in fact often associated with arguments for a "feminine science") is fundamentally antithetical to any meaning that the scientific endeavor might have. To ask women scientists to accept the notion of a different science representing a different reality (as distinct from difference in science) would be to ask them to give up their identity as scientists—in much the same way, incidentally, that traditional science has asked them to give up their identity as women. It is, finally, not so much to counterpose an alternative science as to reinforce the traditional opposition between women and science.

The celebration of difference within science seems therefore to constitute a clear advance over both the monolithic view of science that threatens to exclude diversity, and the dualistic (or relativistic) view that threatens to deny particular meaning to the category, science. At the same time, however, the attempt to avoid the problem of duality by ignoring gender altogether carries within itself a critically undermining flaw: It blocks our perception of the very important ways in which gender has been, and remains, constitutively operative in science. In particular, it ignores the fact that it is precisely in the name of gender that the very diversity we would now like to see celebrated has historically been (and continues to be) excluded. Above all, it ignores the uses of gender in maintaining a monolithic ordering of power.

Just as the engendering of culture in general has shown itself as a way of ordering the power structures of our social and political worlds, the engendering of knowledge, and of scientific knowledge in particular, has served to order the sphere of epistemic power. Knowledge *is* power—in many senses of the term. With the rise of modern science, knowledge came to be understood as a particular kind of power—namely, as the power to dominate nature. In this history, we can see the construction of gender *as* the construction of exclusion—of women, of what is labelled feminine, and simultaneously, of the alternative meanings of power that knowledge might engender. As I've argued in my book on this subject (Keller 1985), the exclusion of the feminine from science has been historically constitutive of a particular definition of science—as incontrovertibly objective, universal, impersonal—and masculine: a definition that serves simultaneously to demarcate masculine from feminine and scientists from non-scientists—even good science from bad. In the past as in the present, the sexual division of emotional and intellectual labor has provided a readily available and much relied upon tool for bolstering the particular claims that science makes to a univocal and hence absolute epistemic authority—not only in the contest between scientists and non-scientists, but equally, in contests internal to science. In turn, of course, the same authority serves to denigrate the entire excluded realm of the feminine—a realm which, as it happens, invariably includes most women.

Given the cultural uses of gender in maintaining a univocal conception of power, any gender-blind advocacy of difference (by men or women, in science or elsewhere) entails some risk. Given its particular uses to exclude those who are the cultural carriers of "femininity" from the apex of epistemic authority, women scientists incur a special risk in ignoring these uses. Once dissociated from gender (and hence from sex), the celebration of difference ironically lends itself to the same ends as the denial of difference—it can serve once again to render women themselves superfluous. The question of values, in this discourse, preempts the question of jobs. Although in many ways philosophically opposed to post-modernism, in one important respect, the response of these scientists converges on a problem that has already become evident in post-modernist literary discourse—a problem to which an increasing number of literary scholars have already begun to call attention. Put simply, the question becomes: when anyone can learn to read like a woman, what need have we for women readers? In other words, difference without gender invites another kind of degeneracy—not quite the denial of difference, but its reduction to indifference, a way of thinking about difference as potentially capable as universality of excluding actual women.

Where advocates of difference within science critically depart from and effectively counter that tendency in post-modernism towards an indefinite proliferation of difference is in their reminder of the constraints imposed by the recalcitrance of nature—their reminder that, despite its ultimate unrepresentability, nature does exist. As feminists, we can offer an equally necessary counter (though necessary for different reasons) to that same proliferation by recalling the constraints imposed by the recalcitrance of sex. In truth—perhaps the one truth we actually do know—neither nature nor sex *can* be named out of existence. Both persist, beyond theory, as humbling reminders of our mortality. The question, of course, is how we can maintain this mindfulness without in the process succumbing to the forces that lend to the names we give to nature and sex the status of reality—the forces that constrain our perception and conception of both nature and sex, first, by their naming as science and gender, and second, by the particular namings of science and gender that are, at any particular time, currently normative.

Our success in maintaining awareness of the bi-polar and dialectical influences of both nature and culture on the categories of gender and science may well in the end depend on the adequacy of our analysis of the nature of the forces that work against such an awareness. If these forces do in fact derive from an underlying contest for power, as the story narrated here suggests, then the most central issue at hand is the relation between gender, science, and power—above all, the uses of particular constructions of gender and science in structuring our conceptual and political landscape of power. As long as power itself remains defined in the unitary terms that have prevailed, the struggles for power that ensue provide fuel, on the one hand, for the collapse between science and nature, and gender with sex, and on the other, for the repudiation of nature and/or sex. In other words, they guarantee the very instability in the concepts of gender and science that continues to plague both feminist and science studies.

Feminist analyses have suggested that it is precisely in the interpenetration of our language of gender and our language of science that the multi-dimensional terrains of nature, of culture, and of power have been transformed into one dimensional contests. If so, the effective defusing of these contests would require a different kind of language, reflecting a higher dimensionality in our landscape—neither homogeneous nor divided, spacious enough to enable multiplicity to survive without degenerating into opposition. In short, we need a language that enables us to conceptually and perceptually negotiate our way between sameness and opposition, that permits the recognition of kinship in difference and of difference among kin; a language that encodes respect for difference, particularity, alterity without repudiating the underlying affinity that is the first prerequisite for knowledge. In this effort, I suggest that the mere fact of sexual difference may itself provide us with one useful reminder: it is, after all, that which simultaneously divides and binds us as a species. But surely, nature, in its mercilessly recalcitrant diversity, provides us with another.

NOTES

Portions of this paper are taken from a talk given at the "Little Three" Conference, Jan. 16, 1986, Amherst, MA.

1. The term "instability" has also been employed by Sandra Harding (1986), but with a rather different charge. Where Harding finds instability productive, both politically and intellectually, I find it—again, both politically and intellectually—an obstacle to productive exchange.

2. See, e.g., the critique of Derrida and Foucault in Teresa de Lauretis, *Technologies of Gender* (Indiana University Press, 1987).

REFERENCES

Harding, Sandra. 1986. The instability of the analytic categories of feminist theory. *Signs* 11 (4):645–664.

Keller, Evelyn Fox. 1983. *A feeling for the organism: The life and work of Barbara McClintock.* New York: W. H. Freeman.

———. 1985. *Reflections on gender and science.* New Haven: Yale University Press.

Koblitz, Ann Hibner. N.d. An historian looks at gender and science. Unpublished manuscript.

Can There Be a Feminist Science?

HELEN E. LONGINO

This paper explores a number of recent proposals regarding "feminist science" and rejects a content-based approach in favor of a process-based approach to characterizing feminist science. Philosophy of science can yield models of scientific reasoning that illuminate the interaction between cultural values and ideology and scientific inquiry. While we can use these models to expose masculine and other forms of bias, we can also use them to defend the introduction of assumptions grounded in feminist political values.

The question of this title conceals multiple ambiguities. Not only do the sciences consist of many distinct fields, but the term "science" can be used to refer to a method of inquiry, a historically changing collection of practices, a body of knowledge, a set of claims, a profession, a set of social groups, etc. And as the sciences are many, so are the scholarly disciplines that seek to understand them: philosophy, history, sociology, anthropology, psychology. Any answer from the perspective of some one of these disciplines will, then, of necessity, be partial. In this essay, I shall be asking about the possibility of theoretical natural science that is feminist and I shall ask from the perspective of a philosopher. Before beginning to develop my answer, however, I want to review some of the questions that could be meant, in order to arrive at the formulation I wish to address.

The question could be interpreted as factual, one to be answered by pointing to what feminists in the sciences are doing and saying: "Yes, and this is what it is." Such a response can be perceived as question-begging, however. Even such a friend of feminism as Stephen Gould dismisses the idea of a distinctively feminist or even female contribution to the sciences. In a generally positive review of Ruth Bleier's book, *Science and Gender,* Gould (1984) brushes aside her connection between women's attitudes and values and the interactionist science she calls for. Scientists (male, of course) are already proceeding with wholist and interactionist research programs. Why, he implied, should women or feminists have any particular, distinctive, contributions to make? There is not masculinist and feminist science, just good

Hypatia vol. 2, no. 3 (Fall 1987). © by Helen E. Longino.

and bad science. The question of a feminist science cannot be settled by pointing, but involves a deeper, subtler investigation.

The deeper question can itself have several meanings. One set of meanings is sociological, the other conceptual. The sociological meaning proceeds as follows. We know what sorts of social conditions make misogynist science possible. The work of Margaret Rossiter (1982) on the history of women scientists in the United States and the work of Kathryn Addelson (1983) on the social structure of professional science detail the relations between a particular social structure for science and the kinds of science produced. What sorts of social conditions would make feminist science possible? This is an important question, one I am not equipped directly to investigate, although what I can investigate is, I believe, relevant to it. This is the second, conceptual, interpretation of the question: what sort of sense does it make to talk about a feminist science? Why is the question itself not an oxymoron, linking, as it does, values and ideological commitment with the idea of impersonal, objective, value-free, inquiry? This is the problem I wish to address in this essay.

The hope for a feminist theoretical natural science has concealed an ambiguity between content and practice. In the content sense the idea of a feminist science involves a number of assumptions and calls a number of visions to mind. Some theorists have written as though a feminist science is one the theories of which encode a particular world view, characterized by complexity, interaction and wholism. Such a science is said to be feminist because it is the expression and valorization of a female sensibility or cognitive temperament. Alternatively, it is claimed that women have certain traits (dispositions to attend to particulars, interactive rather than individualist and controlling social attitudes and behaviors) that enable them to understand the true character of natural processes (which are complex and interactive).[1] While proponents of this interactionist view see it as an improvement over most contemporary science, it has also been branded as soft—misdescribed as non-mathematical. Women in the sciences who feel they are being asked to do not better science, but inferior science, have responded angrily to this characterization of feminist science, thinking that it is simply new clothing for the old idea that women can't do science. I think that the interactionist view can be defended against this response, although that requires rescuing it from some of its proponents as well. However, I also think that the characterization of feminist science as the expression of a distinctive female cognitive temperament has other drawbacks. It first conflates feminine with feminist. While it is important to reject the traditional derogation of the virtues assigned to women, it is also important to remember that women are *constructed* to occupy positions of social subordinates. We should not uncritically embrace the feminine.

This characterization of feminist science is also a version of recently propounded notions of a 'women's standpoint' or a 'feminist standpoint' and suffers from the same suspect universalization that these ideas suffer from. If there is one such standpoint, there are many: as Maria Lugones and Elizabeth Spelman spell out in their tellingly entitled article, "Have We Got a Theory for You: Feminist Theory, Cultural Imperialism, and the Demand for 'The Woman's Voice,' " women are too diverse in our experiences to generate a single cognitive framework (Lugones and Spelman 1983). In addition, the sciences are themselves too diverse for me to think that they might be equally transformed by such a framework. To reject this concept of a feminist science, however, is not to disengage science from feminism. I want to suggest that we focus on science as practice rather than content, as process rather than product; hence, not on feminist science, but on doing science as a feminist.

The doing of science involves many practices: how one structures a laboratory (hierarchically or collectively), how one relates to other scientists (competitively or cooperatively), how and whether one engages in political struggles over affirmative action. It extends also to intellectual practices, to the activities of scientific inquiry, such as observation and reasoning. Can there be a feminist scientific inquiry? This possibility is seen to be problematic against the background of certain standard presuppositions about science. The claim that there could be a feminist science in the sense of an intellectual practice is either nonsense because oxymoronic as suggested above or the claim is interpreted to mean that established science (science as done and dominated by men) is wrong about the world. Feminist science in this latter interpretation is presented as correcting the errors of masculine, standard science and as revealing the truth that is hidden by masculine 'bad' science, as taking the sex out of science.

Both of these interpretations involve the rejection of one approach as incorrect and the embracing of the other as the way to a truer understanding of the natural world. Both trade one absolutism for another. Each is a side of the same coin, and that coin, I think, is the idea of a value-free science. This is the idea that scientific methodology guarantees the independence of scientific inquiry from values of value-related considerations. A science or a scientific research program informed by values is *ipso facto* "bad science." "Good science" is inquiry protected by methodology from values and ideology. This same idea underlies Gould's response to Bleier, so it bears closer scrutiny. In the pages that follow, I shall examine the idea of value-free science and then apply the results of that examination to the idea of feminist scientific inquiry.

II

I distinguish two kinds of values relevant to the sciences. Constitutive values, internal to the sciences, are the source of the rules determining what constitutes acceptable scientific practice or scientific method. The personal, social and cultural values, those group or individual preferences about what ought to be I call contextual values, to indicate that they belong to the social and cultural context in which science is done (Longino 1983c). The traditional interpretation of the value-freedom of modern natural science amounts to a claim that its constitutive and contextual features are clearly distinct from and independent of one another, that contextual values play no role in the inner workings of scientific inquiry, in reasoning and observation. I shall argue that this construal of the distinction cannot be maintained.

There are several ways to develop such an argument. One scholar is fond of inviting her audience to visit any science library and peruse the titles on the shelves. Observe how subservient to social and cultural interests are the inquiries represented by the book titles alone! Her listeners would soon abandon their ideas about the value-neutrality of the sciences, she suggests. This exercise may indeed show the influence of external, contextual considerations on what research gets done/supported (i.e., on problem selection). It does not show that such considerations affect reasoning or hypothesis acceptance. The latter would require detailed investigation of particular cases or a general conceptual argument. The conceptual arguments involve developing some version of what is known in philosophy of science as the underdetermination thesis, i.e., the thesis that a theory is always underdetermined by the evidence adduced in its support, with the consequence that different or incompatible theories are supported by or at least compatible with the same body of evidence. I shall sketch a version of the argument that appeals to features of scientific inference.

One of the rocks on which the logical positivist program foundered was the distinction between theoretical and observational language. Theoretical statements contain, as fundamental descriptive terms, terms that do not occur in the description of data. Thus, hypotheses in particle physics contain terms like "electron," "pion," "muon," "electron spin," etc. The evidence for a hypothesis such as "A pion decays sequentially into a muon, then a positron" is obviously not direct observations of pions, muons and positrons, but consists largely in photographs taken in large and complex experimental apparati: accelerators, cloud chambers, bubble chambers. The photographs show all sorts of squiggly lines and spirals. Evidence for the hypotheses of particle physics is presented as statements that describe these photographs. Eventually, of course, particle physicists point to a spot on a photograph and say things like "Here a neutrino hits a neutron." Such an assertion, however, is an interpretive achievement which involves collapsing theoretical and observational moments. A skeptic would have to be supplied a complicated argument linking the elements of the photograph to traces left by particles and these to particles themselves. What counts as theory and what as data in a pragmatic sense change over time, as some ideas and experimental procedures come to be securely embedded in a particular framework and others take their place on the horizons. As the history of physics shows, however, secure embeddedness is no guarantee against overthrow.

Logical positivists and their successors hoped to model scientific inference formally. Evidence for hypotheses, data, were to be represented as logical consequences of hypotheses. When we try to map this logical structure onto the sciences, however, we find that hypotheses are, for the most part, not just generalizations of data statements. The links between data and theory, therefore, cannot be adequately represented as formal or syntactic, but are established by means of assumptions that make or imply substantive claims about the field over which one theorizes. Theories are confirmed via the confirmation of their constituent hypotheses, so the confirmation of hypotheses and theories is relative to the assumptions relied upon in asserting the evidential connection. Conformation of such assumptions, which are often unarticulated, is itself subject to similar relativization. And it is these assumptions that can be the vehicle for the involvement of considerations motivated primarily by contextual values (Longino 1979, 1983a).

The point of this extremely telescoped argument is that one can't give an a priori specification of confirmation that effectively eliminates the role of value-laden assumptions in legitimate scientific inquiry without eliminating auxiliary hypotheses (assumptions) altogether. This is not to say that all scientific reasoning involves value-related assumptions. Sometimes auxiliary assumptions will be supported by mundane inductive reasoning. But sometimes they will not be. In any given case, they may be metaphysical in character; they may be untestable with present investigative techniques; they may be rooted in contextual, value-related considerations. If, however, there is no a priori way to eliminate such assumptions from evidential reasoning generally, and, hence, no way to rule out value-laden assumptions, then there is no formal basis for arguing that an inference mediated by contextual values is thereby bad science.

A comparable point is made by some historians investigating the origins of modern science. James Jacob (1977) and Margaret Jacob (1976) have, in a series of articles and books, argued that the adoption of conceptions of matter by 17th century scientists like Robert Boyle was inextricably intertwined with political considerations. Conceptions of matter provided the foundation on which physical theories were developed and Boyle's science, regardless of his reasons for it, has been fruitful in ways that far exceed his imaginings. If the presence of contextual influences were grounds for disallowing a line of inquiry, then early modern science would not have gotten off the ground.

The conclusion of this line of argument is that constitutive values conceived as epistemological (i.e., truth-seeking) are not adequate to screen out the influence of contextual values in the very structuring of scientific knowledge. Now the ways in which contextual values do, if they do, influence this structuring and interact, if they do, with constitutive values has to be determined separately for different theories and fields of science. But this argument, if it's sound, tells us that this sort of inquiry is perfectly respectable and involves no shady assumptions or unargued intuitively based rejections of positivism. It also opens the possibility that one can make explicit value commitments and still do "good" science. The conceptual argument doesn't show that all science is value-laden (as opposed to metaphysics-laden)—that must be established on a case-by-case basis, using the tools not just of logic and philosophy but of history and sociology as well. It does show that not all science is value-free and, more importantly, that it is not necessarily in the nature of science to be value-free. If we reject that idea we're in a better position to talk about the possibilities of feminist science.

III

In earlier articles (Longino 1981, 1983b; Longino and Doell 1983), I've used similar considerations to argue that scientific objectivity has to be reconceived as a function of the communal structure of scientific inquiry rather than as a property of individual scientists. I've then used these notions about scientific methodology to show that science displaying masculine bias is not *ipso facto* improper or 'bad' science; that the fabric of science can neither rule out the expression of bias nor legitimate it. So I've argued that both the expression of masculine bias in the sciences and feminist criticism of research exhibiting that bias are—shall we say—business as usual; that scientific inquiry should be expected to display the deep metaphysical and normative commitments of the culture in which it flourishes; and finally that criticism of the deep assumptions that guide scientific reasoning about data is a proper part of science.

The argument I've just offered about the idea of a value-free science is similar in spirit to those earlier arguments. I think it makes it possible to see these questions from a slightly different angle.

There is a tradition of viewing scientific inquiry as somehow inexorable. This involves supposing that the phenomena of the natural world are fixed in determinate relations with each other, that these relations can be known and formulated in a consistent and unified way. This is not the old "unified science" idea of the logical positivists, with its privileging of physics. In its "unexplicated" or "pre-analytic" state, it is simply the idea that there is one consistent, integrated or coherent, true theoretical treatment of all natural phenomena. (The indeterminacy principle of quantum physics is restricted to our understanding of the behavior of certain particles which themselves underlie the fixities of the natural world. Stochastic theories reveal fixities, but fixities among ensembles rather than fixed relations among individual objects or events.) The scientific inquirer's job is to discover those fixed relations. Just as the task of Plato's philosophers was to discover the fixed relations among forms and the task of Galileo's scientists was to discover the laws written in the language of the grand book of nature, geometry, so the scientist's task in this tradition remains the discovery of fixed relations however conceived. These ideas are part of the realist tradition in the philosophy of science.

It's no longer possible, in a century that has seen the splintering of the scientific disciplines, to give such a unified description of the objects of inquiry. But the belief that the job is to discover fixed relations of some sort, and that the application of observation, experiment and reason leads ineluctably to unifiable, if not unified, knowledge of an independent reality, is still with us. It is evidenced most clearly in two features of scientific rhetoric: the use of the passive voice as in "it is concluded that . . . " or "it has been discovered that . . . " and the attribution of agency to the data, as in "the data suggest. . . . " Such language has been criticized for the abdication of responsibility it indicates. Even more, the scientific inquirer, and we with her, become passive observers, victims of the truth. The idea of a value-free science is integral to this view of scientific inquiry. And if we reject that idea we can also reject our roles as passive onlookers, helpless to affect the course of knowledge.

Let me develop this point somewhat more concretely and autobiographically. Biologist Ruth Doell and I have been examining studies in three areas of research on the influence of sex hormones on human behavior and cognitive performance: research on the influence of pre-natal, *in utero,* exposure to higher or lower than normal levels of androgens and estrogens on so-called 'gender-role' behavior in children, influence of androgens (pre- and post-natal) on homosexuality in women, and influence of lower than normal (for men) levels of androgen at puberty on spatial abilities (Doell and Longino, forthcoming).

The studies we looked at are vulnerable to criticism of their data and their observation methodologies. They also show clear evidence of androcentric bias—in the assumption that there are just two sexes and two genders (us and them), in the designation of appropriate and inappropriate behaviors for male and female children, in the caricature of lesbianism, in the assumption of male mathematical superiority. We did not find, however, that these assumptions mediated the inferences from data to theory that we found objectionable. These sexist assumptions did affect the way the data were described. What mediated the inferences from the alleged data (i.e., what functioned as auxiliary hypotheses or what provided auxiliary hypotheses) was what we called the linear model—the assumption that there is a direct one-way causal relationship between pre- or post-natal hormone levels and later behavior or cognitive performance. To put it crudely, fetal gonadal hormones organize the brain at critical periods of development. The organism is thereby disposed to respond in a range of ways to a range of environmental stimuli. The assumption of unidirectional programming is supposedly supported by the finding of such a relationship in other mammals; in particular, by experiments demonstrating the dependence of sexual behaviors—mounting and lordosis—on peri-natal hormone exposure and the finding of effects of sex hormones on the development of rodent brains. To bring it to bear on humans is to ignore, among other things, some important differences between human brains and those of other species. It also implies a willingness to regard humans in a particular way—to see us as produced by factors over which we have no control. Not only are we, as scientists, victims of the truth, but we are the prisoners of our physiology.[2] In the name of extending an explanatory model, human capacities for self-knowledge, self-reflection, self-determination are eliminated from any role in human action (at least in the behaviors studied).

Doell and I have therefore argued for the replacement of that linear model of the role of the brain in behavior by one of much greater complexity that includes physiological, environmental, historical and psychological elements. Such a model allows not only for the interaction of physiological and environmental factors but also for the interaction of these with a continuously self-modifying, self-representational (and self-organizing) central processing system. In contemporary neurobiology, the closest model is that being developed in the group selectionist approach to higher brain function of Gerald Edelman and other researchers (Edelman and Mountcastle 1978). We argue that a model of at least that degree of complexity is necessary to account for the human behaviors studies in the sex hormones and behavior research and that if gonadal hormones function at all at these levels, they will probably be found at most to facilitate or inhibit neural processing in general. The strategy we take in our argument is to show that the degree of intentionality involved in the behaviors in question is greater than is presupposed by the hormonal influence researchers and to argue that this degree of intentionality implicates the higher brain processes.

To this point Ruth Doell and I agree. I want to go further and describe what we've done from the perspective of the above philosophical discussion of scientific methodology.

Abandoning my polemical mood for a more reflective one, I want to say that, in the end, commitment to one or another model is strongly influenced by values or other contextual features. The models themselves determine the relevance and interpretation of data. The linear or complex models are not in turn independently or conclusively supported by data. I doubt for instance that value-free inquiry will reveal the efficacy or inefficacy of intentional states or of physiological factors like hormone exposure in human action. I think instead that a research program in neuro-science that assumes the linear model and sex-gender dualism will show the influence of hormone exposure on gender-role behavior. And I think that a research program in neuroscience and psychology proceeding on the assumption that humans do possess the capacities for self-consciousness, self-reflection, and self-determination, and which then asks how the structure of the human brain and nervous system enables the expression of these capacities, will reveal the efficacy of intentional states (understood as very complex sorts of brain states).

While this latter assumption does not itself contain normative terms, I think that the decision to adopt it is motivated by value-laden considerations—by the desire to understand ourselves and others as self-determining (at least some of the time), that is, as capable of acting on the basis of concepts or representations of ourselves and the world in which we act. (Such representations are not necessarily correct, they are surely mediated by our cultures; all we wish to claim is that they are efficacious.) I think further that this desire on Ruth Doell's and my part is, in several ways, an aspect of our feminism. Our preference for a neurobiological model that allows for agency, for the efficacy of intentionality is partly a validation of our (and everyone's) subjective experience of thought, deliberation, and choice. One of the tenets of feminist research is the valorization of subjective experience, and so our preference in this regard conforms to feminist research patterns. There is, however, a more direct way in which our feminism is expressed in this preference. Feminism is many things to many people, but it is at its core in part about the expansion of human potentiality. When feminists talk of breaking out and do break out of socially prescribed sex-roles, when feminists criticize the institutions of domination, we are thereby insisting on the capacity of humans—male and female—to act on perceptions of self and society and to act to bring about changes in self and society on the basis of those perceptions. (Not overnight and not by a mere act of will. The point is that we act.) And so our criticism of theories of the hormonal influence or determination of so-called gender-role behavior is not just a rejection of the sexist bias in the description of the phenomena—the behavior of the children studied, the sexual lives of lesbians, etc.—but of the limitations on human capacity imposed by the analytic model underlying such research.[3]

While the argument strategy we adopt against the linear model rests on a certain understanding of intention, the values motivating our adoption of that understanding remain hidden in that polemical context. Our political commitments, however, presuppose a certain understanding of human action, so that when faced with a conflict between these commitments and a particular model of brain-behavior relationships we allow the political commitments to guide the choice.

The relevance of my argument about value-free science should be becoming clear. Feminists—in and out of science—often condemn masculine bias in the sciences from the vantage point of commitment to a value-free science.

Androcentric bias, once identified, can then be seen as a violation of the rules, as "bad" science. Feminist science, by contrast, can eliminate that bias and produce better, good, more true or gender free science. From that perspective the process I've just described is anathema. But if scientific methods generated by constitutive values cannot guarantee independence from contextual values, then that approach to sexist science won't work. We cannot restrict ourselves simply to the elimination of bias, but must expand our scope to include the detection of limiting and interpretive frameworks and the finding or construction of more appropriate frameworks. We need not, indeed should not, wait for such a framework to emerge from the data. In waiting, if my argument is correct, we run the danger of working unconsciously with assumptions still laden with values from the context we seek to change. Instead of remaining passive with respect to the data and what the data suggest, we can acknowledge our ability to affect the course of knowledge and fashion or favor research programs that are consistent with the values and commitments we express in the rest of our lives. From this perspective, the idea of a value-free science is not just empty, but pernicious.

Accepting the relevance to our practice as scientists of our political commitments does not imply simple and crude impositions of those ideas onto the corner of the natural world under study. If we recognize, however, that knowledge is shaped by the assumptions, values and interests of a culture and that, within limits, one can choose one's culture, then it's clear that as scientists/theorists we have a choice. We can continue to do establishment science, comfortably wrapped in the myths of scientific rhetoric or we can alter our intellectual allegiances. While remaining committed to an abstract goal of understanding, we can choose to whom, socially and politically, we are accountable in our pursuit of that goal. In particular we can choose between being accountable to the traditional establishment or to our political comrades.

Such accountability does not demand a radical break with the science one has learned and practiced. The development of a "new" science involves a more dialectical evolution and more continuity with established science than the familiar language of scientific revolutions implies.

In focusing on accountability and choice, this conception of feminist science differs from those that proceed from the assumption of a congruence between certain models of natural processes and women's inherent modes of understanding.[4] I am arguing instead for the deliberate and active choice of an interpretive model and for the legitimacy of basing that choice on political considerations in this case. Obviously model choice is also constrained by (what we know of) reality, that is, by the data. But reality (what we know of it) is, I have already argued, inadequate to uniquely determine model choice. The feminist theorists mentioned above have focused on the relation between the content of a theory and female values or experiences, in particular on the perceived congruence between interactionist, wholist visions of nature and a form of understanding and set of values widely attributed to women. In contrast, I am suggesting that a feminist scientific practice admits political considerations as relevant constraints on reasoning, which, through their influence on reasoning and interpretation, shape content. In this specific case, those considerations in combination with the phenomena support an explanatory model that is highly interactionist, highly complex. This argument is so far, however, neutral on the issue of whether an interactionist and complex account of natural processes will always be the preferred one. If it is preferred, however, this will be because of explicitly political considerations and not because interactionism is the expression of "women's nature."

The integration of a political commitment with scientific work will be expressed differently in different fields. In some, such as the complex of research programs having a bearing on the understanding of human behavior, certain moves, such as the one described above, seem quite obvious. In others it may not be clear how to express an alternate set of values in inquiry, or what values would be appropriate. The first step, however, is to abandon the idea that scrutiny of the data yields a seamless web of knowledge. The second is to think through a particular field and try to understand just what its unstated and fundamental assumptions are and how they influence the course of inquiry. Knowing something of the history of a field is necessary to this process, as is continued conversation with other feminists.

The feminist interventions I imagine will be local (i.e., specific to a particular area of research); they may not be exclusive (i.e., different feminist perspectives may be represented in theorizing); and they will be in some way continuous with existing scientific work. The accretion of such interventions, of science done by feminists as feminists, and by members of other disenfranchised groups, has the potential, nevertheless, ultimately to transform the character of scientific discourse.

Doing science differently requires more than just the will to do so and it would be disingenuous to pretend that our philosophies of science are the only barrier. Scientific inquiry takes place in a social, political and economic context which imposes a variety of institutional obstacles to innovation, let alone to the intellectual working out of oppositional and political commitments. The nature of university career ladders means that one's work must be recognized as meeting certain standards of quality in order that one be able to continue it. If those standards are intimately bound up with values and assumptions one rejects, incomprehension rather than conversion is likely. Success requires that we present our work in a way that satisfies those standards and it is easier to do work that looks just like work known to satisfy them than to strike out in a new direction. Another push to conformity comes from the structure of support for science. Many of the scientific ideas argued to be consistent with a feminist politics have a distinctively non-production orientation.[5] In the example discussed above, thinking of the brain as hormonally programmed makes intervention and control more likely than does thinking of it as a self-organizing complexly interactive system. The doing of science, however, requires financial support and those who provide that support are increasingly industry and the military. As might be expected they support research projects likely to meet their needs, projects which promise even greater possibilities for intervention in and manipulation of natural processes. Our sciences are being harnessed to the making of money and the waging of war. The possibility of alternate understandings of the natural world is irrelevant to a culture driven by those interests. To do feminist science we must change the social and political context in which science is done.

So: can there be a feminist science? If this means: is it in principle possible to do science as a feminist?, the answer must be: yes. If this means: can we in practice do science as feminists?, the answer must be: not until we change present conditions.

NOTES

I am grateful to the Wellesley Center for Research on Women for the Mellon Scholarship during which I worked on the ideas in this essay. I am also grateful to audiences at UC Berkeley, Northeastern University, Brandeis University and Rice University for their comments and to the anonymous reviewers for *Hypatia* for their suggestions. An earlier version appeared as Wellesley Center for Research on Women Working Paper #63.

1. This seems to be suggested in Bleier (1984), Rose (1983) and in Sandra Harding's (1980) early work.
2. For a striking expression of this point of view see Witelson (1985).
3. Ideological commitments other than feminist ones may lead to the same assumptions and the variety of feminisms means that feminist commitments can lead to different and incompatible assumptions.
4. Cf. note 1, above.
5. This is not to say that interactionist ideas may not be applied in productive contexts, but that, unlike linear causal models, they are several steps away from the manipulation of natural processes immediately suggested by the latter. See Keller (1985), especially Chapter 10.

REFERENCES

Addelson, Kathryn Pine. 1983. The man of professional wisdom. In *Discovering reality*, ed. Sandra Harding and Merrill Hintikka. Dordrecht: Reidel.

Bleier, Ruth. 1984. *Science and gender.* Elmsford, NY: Pergamon.

Doell, Ruth, and Helen E. Longino. N.d. *Journal of Homosexuality.* Forthcoming.

Edelman, Gerald and Vernon Mountcastle. 1978. *The mindful brain.* Cambridge, MA: MIT Press.

Gould, Stephen J. 1984. Review of Ruth Bleier, *Science and gender. New York Times Book Review,* VVI, 7 (August 12): 1.

Harding, Sandra. 1980. The norms of inquiry and masculine experience. In *PSA 1980,* Vol. 2, ed. Peter Asquith and Ronald Giere. East Lansing, MI: Philosophy of Science Association.

Jacob, James R. 1977. *Robert Boyle and the English Revolution, A study in social and intellectual change.* New York: Franklin.

Jacob, Margaret C. 1976. *The Newtonians and the English Revolution, 1689–1720.* Ithaca, NY: Cornell University Press.

Keller, Evelyn Fox. 1985. *Reflections on gender and science.* New Haven, CT: Yale University Press.

Longino, Helen. 1979. Evidence and hypothesis. *Philosophy of Science* 46 (1): 35–56.

———. 1981. Scientific objectivity and feminist theorizing. *Liberal Education* 67 (3): 33–41.

———. 1983a. The idea of a value free science. Paper presented to the Pacific Division of the American Philosophical Association, March 25, Berkeley, CA.

———. 1983b. Scientific objectivity and logics of science. *Inquiry* 26 (1): 85–106.

———. 1983c. Beyond "bad science." *Science, Technology and Human Values* 8 (1): 7–17.

Longino, Helen and Ruth Doell. 1983. Body, bias and behavior. *Signs* 9 (2): 206–227.

Lugones, Maria and Elizabeth Spelman. 1983. Have we got a theory for you! Feminist theory, cultural imperialism and the demand for "the woman's voice." *Hypatia 1,* published as a special issue of *Women's Studies International Forum* 6 (6): 573–581.

Rose, Hilary. 1983. Hand, brain, and heart: A feminist epistemology for the natural sciences. *Signs* 9 (1): 73–90.

Rossiter, Margaret. 1982. *Women scientists in America: Struggles and strategies to 1940.* Baltimore, MD: Johns Hopkins University Press.

Witelson, Sandra. 1985. An exchange on gender. *New York Review of Books* (October 24).

Is the Subject of Science Sexed?

LUCE IRIGARAY
Translated by Carol Mastrangelo Bové

The premise of this paper is that the language of science, like language in general, is neither asexual nor neutral. The essay demonstrates the various ways in which the non-neutrality of the subject of science is expressed and proposes that there is a need to analyze the laws that determine the acceptability of language and discourse in order to interpret their connection to a sexed logic. C.B.

How do you speak to scientists? More important, scientists in different disciplines, each forming a world, each system of each of these worlds intending to be global at a given moment? If, at each instant, each of these worlds is organized in a way that is total, closed, how can you reopen these worlds to have them meet, to have them speak to each other? In what language? According to what mode of discourse?

The question seems insoluble. Every scientific universe seems to have its world view, its stakes, its protocols for action, its techniques, its syntax. It seems to be isolated, cut off from others. From what point of view, then, do you *survey* these different horizons to find meeting places, feasible intersections, possible movement among them? *What right do you have to an outside point of reference?* How do you obtain it? Historically, there has been God, transcending every *épistémé.* But if, when "science is in power, God is dead" (Nietzsche), from where do you assemble these worlds? My hypothesis, then, will be that the place for common inquiry is *inside* and not *outside,* underlying and not simply transcendent, "underground" and not only "in the sky," very deeply buried and not relegated to some unquestionable absolute guarantee.

How do you find this possible place for inquiry and make it visible? *How can we talk about it?* Neither *I*, nor *you,* nor *we* appears in the language of science. The *subjective* is forbidden there unless you go to the more or less secondary sciences, the *human sciences* of which we continue to ask if they are completely sciences, substitutes, for literature, for poetry. Or again, if they are true or false, verifiable or falsifiable, formalizable or always ambiguous because expressed in natural language, too empirical or too metaphysical,

Hypatia vol. 2, no. 3 (Fall 1987).

deriving from the axiomatization of the so-called exact sciences or resisting this formalization, etc. These are old debates, old quarrels, that are always relevant, with eventual reversals of power, falls and rebirths of imperialism.

These cycles can reproduce themselves indefinitely. But maybe we can ask whether there is not, underlying, underground, a *common producer* that makes science. But who? Is there someone? Does he show himself? How do we summon him? It has been a long time since I have experienced so much difficulty at the idea of speaking in public. Usually I can anticipate whom I'm going to speak to, how to speak, how to argue, to have myself understood, to plead my case, even to please or to displease. This time I know nothing, because I do not know whom I have in front of me. The reverse of scientific imperialism: not knowing whom you are addressing, how do you speak? The anxiety of an absolute power that floats through the air, of an authoritative judgment everywhere invisibly present, of a tribunal that at its limit has neither judge, nor lawyer nor accused. But the legal system is in place. There is a truth to which you have to submit without appeal, against which you can transgress without wanting to, without knowing it. This supreme instance acts upon your unwilling body. No one is responsible for this terror, for this terrorism. Yet they function. In any case, in this classroom for presentations and lectures. For me at any rate. If I met each of you separately, I think I would find a way to say *you, I, we*. But here, in the name of science?

My first question would be: *what schiz does science produce* in the person who practices it or mobilizes it in one way or another? What desire (individual or social) is at stake in him or her when they do science, what other desire when they make love? What schiz and *what explosion*: pure science on the one hand, politics on the other, nature or art on a third or as conditions of possibilities, love on a fourth? This schiz, this explosion, perhaps claimed by you as an exit from scientific imperialism, are they not already programmed by it as separation of the subject from himself, from his desires, but still dispersion in multiple sectors, including science, among which sectors meetings become impossible, verification of responsibilities unfeasible? There remains an imperialistic *there is* or a *they* whose delegates to power, whose politics perform as the opportunity arises. At the moment when scientists react, isn't the die already cast in the name of science? Imperialism without a subject.

So, to take a different handle on things and to play a little in this genre of questionnaire that flourishes in women's magazines (replacing the crossword puzzles of the daily newspapers that are too often exclusively male?), let's try:

- If I tell you that two eggs can produce a new being, does this discovery seem to you possible, probable, true? Purely genetic? Or also of an order that is social, economic, cultural, political? Belonging to the exact sciences? Put a check there or in the appropriate boxes. Will this type of discovery be encouraged and subsidized? Will it be disseminated by the media? Yes? No? Why?

Answer? How do you interpret the answer? By the importance of sperm in patriarchy and its link with the standard of goods and the Symbolic? By the importance of reproduction in its ambiguous correlation to pleasure and to desire in the differences between the sexes?

And, as long as we're on this question of reproduction and of its hormonal constituents:

- Is masculine contraception hormonally possible? Yes? No? Why? If it is, is this information disseminated, is its practice encouraged?

- Is the left hemisphere less developed in women than in men? Yes? No? Is this discovery used to justify the social, cultural, and political inferiority of women? Does this observation talk about something innate or acquired? Give your interpretation according to your hypothesis. Indicate what connection you make with the inhabitants of certain Oriental countries that, science tells us, share the same anatomical destiny. Does the mode of physical and mental practices that originates in these Asiatic countries signify an unconscious (?) desire on the part of men to become women? A resistance to women's liberation by means of an appropriation of all values, of a failure to recognize a "sexed" symbolic morphology?

- The female child, according to a good number of observations, enjoys a more precocious development than the boy child: she speaks more quickly and her social skills are more advanced. Yes? No? Verifiable? Falsifiable? Does she use these skills to become a desirable object for others? Hence her regression? True? False? Justify your answer.

- What is the percentage of men and women in the world's population? What is the percentage of men and women in the political, social, and cultural government of this population? Does this go without saying? Does it correspond to a masculine or feminine *nature*, to man's or woman's desire? Is it innate or acquired?

- Are women *naturally* more limited and ignorant, more animal or capable of language than men? Are they more inept at political, economic, social, cultural leadership? Innate? Acquired? Verifiable? Falsifiable?

- Is a woman who does science a man with full privileges? A genetic aberration? A monster? A bisexual human? A submissive or an unsubmissive woman? Or . . . ?

- Is there or is there not a dominant discourse that claims to be universal and neutral from the point of view of the difference between the sexes? Do you agree to perpetuate it? One year, two years, one hundred years or always?

- Who, according to our epistemological tradition, is the keystone of the order of discourse?

- Why has God always been and why is he still, in the West at least, God the *father*? That is, the exclusively *masculine* pole of sexual difference? Is this to designate in this way the sex that is hidden in and beyond all discourse? Or . . . ?

In fact, what claims to be universal is the equivalent of a male idiolect, a masculine Imaginary, a *sexed* world—without neutrality. Unless you claim to be an unbridled defender of idealism, there is nothing surprising in this. It has always been men who have spoken and especially written, in the sciences, philosophy, religion, politics.

But nothing is said about the *intuition* of the scholar. It is as if it were produced *ex nihlo.* Yet certain modalities or qualities of this intuition can be pinpointed. It is always a question of:

- positing *a* world in front of you, constituting a world *in front of yourself*;
- imposing a *model* on the universe to appropriate it, an imperceptible, invisible model *projected* like a piece of clothing that encompasses it. Isn't this to dress it blindly in your identity?
- claiming to be rigorously foreign to the model, proving that the model is purely and simply *objective*;
- demonstrating the *imperceptibility* of the model while it is always prescribed at least by the privilege of the *visible,* by the absence, the distancing of a subject nevertheless surreptitiously there;
- an imperceptibility that is possible thanks to the *instrument's mediation,* to the intervention of a technique that separates the subject from its object of investigation, to the process of distancing and of delegating power to what comes between the observed universe and the observing subject;
- constructing an "*ideal*" model, independent of the physical and psychical models of the producer, according to the games of induction and deduction and always passing through an ideal formulation;
- proving the *universality* of the model, at least in time x, and its absolute power (independent of the producer), its constitution of a unique and total world;
- laying out this *universality* by means of protocols for action that at least two (identical?) subjects agree on;
- proving that the discovery is *efficacious, productive, profitable, exploitable (exploiting?* in a more or less inanimate way?)—meaning *progress.*

These characteristics reveal an isomorphism in man's sexual Imaginary, an isomorphism which must remain rigorously masked. "Our subjective experiences and our beliefs can never justify any utterance," affirms the epistemologist of the sciences.

You must add that all of these discoveries must be expressed in a language that is *well-written,* meaning *reasonable,* that is:

- expressed in symbols or letters, interchangeable with *proper nouns,* that refer only to an intratheoretical object, thus to no character or object from the real or from reality. The scholar enters into a fictional universe that is incomprehensible to those who do not participate in it.

- the formative signs for terms and for predicates are:

+ : or definition of a new term

= : which indicates a property by equivalence and substitution (belonging to a whole or to a world);

ϵ: signifying belonging to an object type

- the *quantifiers* (and not *qualifers*) are:

>> <<;

the universal quantifier;

the existential quantifier submitted, as its name indicates, to the quantitative.

According to the semantics of incomplete beings (Frege), functional symbols are variables found at the boundary of the identity of syntactic forms and the dominant role is given to the universality symbol or universal quantifier.

- the *connectors* are:
- negation: P or not P;
- conjunction: P or Q;
- disjunction: P or Q;
- implication: P implies Q:
- equivalence: P equals Q;

There is then no sign:

- of *difference* other than the quantitative;
- of *reciprocity* (other than within a common property or a common whole)
- of *exchange*;
- of *permeability*;
- of *fluidity*.

Syntax is governed by:

- *identity with,* expressed by property and quantity;
- *non-contradiction* with ambiguity, ambivalence, polyvalence minimized;
- *binary oppositions*; nature/reason, subject/object, matter/energy, inertia/movement.

It is clear that formal language simply corresponds to the rules of the game. This language serves to define the game in such a way that a decision is possible in case of a disputed move. But who are the participants? How do you have an intuition outside of the language employed? If you have one, how do you translate it for the participants?

But the non-neutrality of the subject of science is expressed in different ways. It can be interpreted through what is uncovered or what is not uncovered at a moment in history, in what science takes or does not take as the stakes of its research. That is, proceeding more or less without order or respect for the hierarchy of the sciences:

- *psychoanalytic "science"* is based upon the first two principles of thermodynamics which underlie Freud's model of the libido. Now these two principles seem more like isomorphs of masculine rather than feminine sexuality. The latter is less subject to the tension-discharge alternation, to the conservation of required energy, to the maintenance of balanced states, to the functioning through saturation of closed and open circuits, to the reversibility of time, etc. Feminine sexuality would perhaps harmonize better, if you need to evoke a scientific model, with what Prigogine calls "dissipative" structures which function by means of an exchange with the exterior world, which proceed by energy levels and whose order is not one that seeks balance but one that seeks passage over thresholds corresponding to a movement beyond disorder or entropy without any discharge.

- the *economic sciences* (and also social?) have emphasized the phenomenon of scarcity and the question of survival more than those of life and surplus.

- the *linguistic sciences* have latched on to models of the *utterance,* to the synchronic structures of language, to the language models of which any normally constituted subject has the intuition. They have not considered and even refuse at times to consider the question of the sexuality of discourse. They allow, by necessity, certain lexical terms to be added to the approved stock, new stylistic figures to impose themselves eventually, but they do not imagine that syntax and syntactic-semantic functioning are sexually determined, and neither neutral nor universal nor atemporal.

- the *biological sciences* approach certain questions very late: that of the constitution of placental tissue, that of membrane permeability, for example. Questions much more directly correlated to the female maternal sexual Imaginary?

- the *mathematical sciences,* in the theory of wholes, concern themselves with closed and open spaces, with the infinitely big and the infinitely small. They concern themselves very little with the question of the partially open, with wholes that are not clearly delineated, with any analysis of the problem of borders, of the passage between, of fluctuations occurring between the thresholds of specific wholes. Even if topology suggests these questions, it emphasizes what closes rather than what resists all circularity.

- the *logical sciences* are more interested in bivalent rather than in trivalent or polyvalent theories that still appear to be marginal.

- the *physical sciences* constitute their object according to a nature that they measure in a way that is more and more formal, abstract, based upon a model. Their techniques show an increasingly sophisticated axiomatization and deal with matter that certainly still exists but that is not visible to the subject performing the experiment, at least in most of the areas of this science. Nature, the stakes of the physical sciences, is in danger of being exploited and disintegrated by the physicist without his even knowing it. The Newtonian break has ushered scientific enterprise into a world where sense perception is worth little, a world which can lead to the annihilation of the very stakes of physics' object: the matter (whatever the predicates) of the universe and of the bodies that constitute it. In this very science, however, cleavages exist: quantum theory/field theory, mechanics of solids/dynamics of fluids, for example. But the imperceptibility of the matter under study often brings with it the paradoxical privilege of *solidity* in discoveries and a delay, even an abandoning of the analysis of the infinity of the fields of force. Could this be interpreted as a result of failing to take into account the dynamics of the subject in search of itself?

Faced with these observations, these questions, does the alternative become *either* do science *or* "be a militant"? Or again, to continue to do science *and* to divide yourself up into different functions, several persons or characters? Should the truth of science and that of life remain separate, at least for the majority of researchers? What science and what life is then under discussion? Especially since life in our time is greatly dominated by science and its techniques.

What is the source of the schiz that is imposed and accepted by scientists? A model of an unanalyzed subject? A "subjective" revolution that has not taken place, the explosion of the subject having been programed by the épistémé and by the power structures that it has put in place? The epistemological subject has undergone but not yet acted upon nor gone beyond the Copernican revolution? The discourse of this subject becomes modified but in an even more disappropriating way than the language about the world that preceded it. The scientist now wants to be *in front of* the world, naming, legislating, axiomatizing it. He manipulates nature, utilizes it, exploits it, but forgets that he is also in nature, that he is still physical and not only in front of the phenomena whose physical nature he fails to recognize. Advancing according to an objective method that would shelter him from all instability, from all moods, from all emotions and affective fluctuations, from all intuitions that are not programed in science's name, from all interference by his desires, especially those that are sexed, he settles himself down, in his discoveries, in the systematic, in what can be assimilated to the already dead? Fearing, sterilizing the imbalances that are in fact needed to arrive at a new frontier.

One of the places most likely to provoke a questioning of the scientific landscape is that of the examination of the subject of science and its psychic and sexed implication in discourse, discoveries and their organization.

In order to ask if so-called universal language and discourse (languages and discourses, at least one of which is that of science) are neutral as far as the sex that produces them, it is appropriate to pursue research according to a double exigency: *to interpret the authoritative discourse* as one that obeys a sexual order that the speaking subject doesn't see and *to try to define the characteristics of what a language that is differently* sexed *would be.*

In other words, is there, within the logical and syntactic/semantic framework of existing discourse, an opening or a degree of freedom that allows the expression of sexual difference? It is, then, a question of analyzing the laws (including those not articulated as such) that determine the acceptability of language and discourse in order to interpret their connection to a sexed logic. This analysis can be pursued from a number of different perspectives:

- the study of the *causal* modalities that dominate current discourse considered normal and the modalities of *the conditional* and *the unreal, the restricted,* etc., which fix the "practicable," limiting the freedom of a speaking subject that would not obey certain criteria of normality. If they allow information to accumulate (a certain kind of already coded communication), do not these causal and restrictive modalities (the two are linked) introduce the intradiscursive brakes that preclude the possibility of a qualitatively different utterance?

- the means or conjunctions of *coordination* seem to participate in this economy where the causality principle dominates a so-called asexed discourse: juxtapositions that extend to and include the addition of propositions and subjects (and . . . and), alternatives (or . . . or), exclusion to the point where a subject of enunciation is eventually eliminated (neither . . . nor), coordination that moves in the direction of the syllogistics that regulates discourse (because, thus, but).

- What modalities of subordination or coordination would authorize a discursive relationship between two sexually different subjects?

- *the analysis of symmetry* (especially right-left) in intersubjective relationships and its impact on language production. Could the question of symmetry and asymmetry give rise to certain criteria for determining a qualitative difference between the sexes? Is the "blind spot of a dream of symmetry"[1] situated in the same place in a relation between the same sexes or different sexes? But this very dream that may underpin the economy of the speaking subject seems to be undercut by universal laws to which no one who studies nature and language (nor any foreign speaker or interlocutor) can remain indifferent.

- if women are kept within a *potential* language, they constitute an energy reserve that can be destroyed or that may explode if possible modes of expression are lacking. If women represent only discourse's other side or reverse (in mirror symmetry?), they turn it back upon itself. Constrained by a *defensive or offensive mimetics* and lacking a possible response, they are in danger of absorbing discourse's meaning by breaking it open. They intercept, then, discourse's teleology or intentionality, an interception that accelerates a process of destructuration that is appropriate if a new language is to have a place. The question is whether women's language, remaining within discourse's same general economy, uses a meaning that is potential and not yet realized or whether what they think and can say demands a transformation of language's frontiers. The latter would explain the resistance to their entry into communication networks and even more into the theoretical and scientific places that determine values and the laws of exchange.

Certain questions can be raised concerning women's access to language and discourse:

- why is their energetic potential for language always at the *vanishing point,* lacking a possible return to the subject of enunciation? Certain recent studies in the theory of discourse, and also in physics, seem able to shed light upon the place of their previously blocked access to discursivity. You need to return to the examination of temporality and of its relationship to the starting point where a subject/producer of language can or cannot be situated. If the discourse of a supposed speaker intercepts speech, cutting her off from a memory of the past and from an anticipation of the future, only attempts to regain the place where she can make herself heard remains for the subject. This fact underlines the importance of the *local* in the constructions of women's language. Circumstances relating to *place* would determine for the most part the programing of the "discourse."

- in this insistence on the question of place, is there not an attempt to give shape once again to a subject of enunciation that lacks temporalization in a dynamics of communication? The question of the possible or impossible *reversibility* of an utterance, especially between speaker, and hearer, should be imagined from this perspective as well as from that of its *repetition* or eventual reproduction. These conditions seem indispensable to an acceptable discourse, the other being positioned as a mirror that at one and the same time inverts the discourse that it receives and responds to it once this inversion takes place.

- the problem of the mirror, possible or impossible for the other, dominates the enigma of women's language and silence. In any case, "women" do not say nothing, and the fascination experienced, especially by certain psychologists, in connection with what women say certainly means that something like a possible deciphering of the production of language expresses itself through them.

These questions could also be approached from the following point of view:

-Does what we call *mother tongue* organize a specific production of language by the mother and of exchange between mother and children? Is not socially acceptable language always paternal? Does this situation open up language at the point where it enters discourse? An opening that continually threatens the latter with collapse, with insanity, with a paralyzing normalization.

- At the origins of our culture the maternal creation of language in all its forms has been blocked. The maternal has become relegated to procreation and is not the place for a productive matricial function. From this perspective, it is appropriate to reexamine, to reinterpret the Freudian text, notably *Totem and Taboo,* which defines the murder of the father and the sons' sharing of his body as actions that give rise to the primitive horde. Would there not be a more archaic matricide beneath the murder of the father, a matricide that could be deciphered at the origins of our culture (in Greek tragedies, mythologies, and even philosophies)? This murder of the mother (as fertile lover in a cultural dimension), would continue to operate in the positioning of the symbolic and social order that is ours. What consequences does this matricide have on the production of language and the programing of discourses (including the scientific)?

- Since psychoanalytic "science" is said to be a theory of the subject, *Freud's hypothesis concerning the constitution of the subject's relationship to discourse* deserves to be reconsidered, reinterpreted. Freud proposes the "bobbin game" as the scene that introduces the subject into language. The child—in this case a boy—would master the mother's absence by using an instrument that he throws far from him, brings back, moves away from or close to his place, in his place, while accompanying this gesture with the emission of alternating vowels: o-o-o (away), a-a-a (close).

This "game" called *fort-da,* including its vowel (o-a) alternations, would mark the child's entrance into the possibility of symbolic distantiation. This entrance would take place by assimilating the mother to an object held by a string thanks to which distance from her would be overcome or destroyed by the boy-child (Freud does not put forward the hypothesis of what would happen for the girl), a gesture accompanied by sound emissions, a kind of musical keyboard. Does this *fort-da* scene continue to retain a significant function in constituting the meaning of language? How are vowels articulated with consonants?

This stage set described by Freud supposes the absence of the mother as speaker, the presence of the grandfather as observer and regulator of a *normal* language. What gestures and language leftovers, especially between child and mother, between mother and child, are in this way abandoned outside of an acceptable discourse? Doesn't this *hors-lieu* of the spoken, of the speakable, bring with it the systematicity and also the insanity of supposedly acceptable discourse, given that a practice of *exchange between* mother and son, mother and subject-man is not in place in language? But for Freudian theory to function, this other means of distantiation could not become one that murders.

- Freud says nothing about the little girl's entrance into language except that she is more precocious than the little boy. But he does not describe her first scene of gestural and verbal symbolization, especially in her relationship to her mother. On the contrary, he affirms that the girl must leave her mother, turn away from her, to enter into desire, the order of the father, of man. An economy of gestural and verbal relations between mother and daughter, between women, would in this way be annulled, abolished, forgotten, by a so-called normal language that is nevertheless neither asexual nor neutral. Does discourse remain at the level of partially theoretical exchanges between generations of men concerning mastery of the mother, of nature? The fertile ground of a sexed speech, of a sexual creation and not simply procreation, is lacking.

NOTES

This essay orginally appeared in *Les Temps Modernes* 9, No. 436 (November, 1982), pp. 960–974. It was reprinted in Luce Irigaray (1985) *Parler, N'est Jamais Neutre* (Paris: Les Editions de Minuit). This is the translation of a revised version of an essay that was first translated in *Cultural critique* (1985), pp. 73–88.

1. See Irigaray's (1985) *Speculum of the Other Woman*, trans. Gillian C. Gill (Ithaca, NY: Cornell University Press). C.B

Uncovering Gynocentric Science

RUTH GINZBERG

Feminist philosophers of science have produced an exciting array of works in the last several years, from critiques of androcentrism in traditional science to theories about what might constitute feminist science. I suggest here another possibility: that gynocentric science has existed all along, then the task of identifying a feminist alternative to androcentric science should be a suitable candidate for empirical investigation. Such empirical investigation could provide a solid ground for further theorizing about feminist science at a time when that solid ground is looking rather necessary.

Recent feminist critiques of science have documented a wide variety of forms of androcentrism in traditional Western science (Griffin 1980; Merchant 1980; Bleier 1984; Keller 1985; Harding 1986; Birke 1986; Bleier 1986). There have been some attempts made to define a gynocentric conception of science; Evelyn Fox Keller (1983, 1985) for example, has suggested that we might find clues about gynocentric science by examining the work of women scientists like Barbara McClintock. But many feminist philosophers of science, including Keller, are still at a loss when asked to define what a truly gynocentric alternative might look like. Ruth Bleier articulates the question that many of us struggle with: "[H]ow can we even begin to conceptualize science as non-masculine . . . when most of written civilization—our history, language, conceptual frameworks, literature—has been generated by men?" (Bleier 1986, 15). Some have suggested that we are not yet in a position to identify a fully articulated feminist successor science (Fee 1983; Harding 1986; Rose 1986). On a Kuhnian model of scientific paradigms, a successor science that would follow the current andro-Eurocentric paradigm could not be fully articulated at this time, partly because of incommensurability and partly because paradigms are never fully articulated even in their own fullest maturity (Kuhn 1970). I would like to toss yet another suggestion into the realm of discourse about feminist science in partial response to Bleier's question: the suggestion that there has been gynocentric science all along,

Hypatia vol. 2, no. 3 (Fall 1987). © by Ruth Ginzberg.

but that we often fail to recognize it as gynocentric *science* because it traditionally has not been awarded the honorific label of 'Science.'

Taking a cue from so many other feminist inquiries, I would like to reexamine women's actual activities in order to discover clues about gynocentric science. My hunch is that if there is such a thing as gynocentric science, it is unlikely that it is just now beginning. I suspect that there is such a thing, and that it has been practiced throughout history—just as other gynocentric traditions have existed throughout history—but that the androcentric recordkeepers have failed to notice or record it. In the same way that feminists are beginning to recover some of our artistic, political, spiritual and social traditions, I believe that we can now recover some of the scientific traditions of our foresisters by reviewing history with a feminist eye.

For a start, it seems important not to confine our review of history to those activities which have been officially labeled 'Science' until now. As Feyerabend (1975) has argued, 'Science' is—at least in part—a political term. If Feyerabend is correct about this, and I am convinced that he is, then it is imperative that we look beyond the 'official' histories to correct for the political factors working against women. As is typical of oppressed groups, much of women's activity has been outside of the mainstream of Western culture. But that doesn't mean that these activities weren't occurring, or that they weren't valuable, nor does it necessarily mean that they weren't science. It only means that they weren't the subject of favorable attention from the members of the dominant cultures. What I am suggesting is that there are women's activities that haven't been called 'Science' for *political* reasons, even when those activities have been model examples of inquiry leading to knowledge of the natural world. So my partial answer to Bleier's question is that we must look outside of the histories, conceptual frameworks, literature, and possibly even the language, that have been generated by men.

The question, then, becomes that of how to begin. I like Keller's approach as a starting point: She has examined carefully the life and work of a woman scientist who *has* been acknowledged for her scientific work, but who was often seen as a bit "odd" or "incomprehensible." These are hallmarks of the sort of paradigmatic incommensurability described by Kuhn (1970), and should serve as clues about the possible existence of another scientific paradigm. In her biography of Barbara McClintock, Keller (1983, 201) points to such things as "a deep reverence for nature, a capacity for union with that which is to be known," and a sort of holism of approach as being thematic in McClintock's work. There are hints of this sort of theme as well in the work of Rachel Carson, who introduced the concept of ecology to the American public. Anticipating some of the recent, more overtly feminist critiques, Carson wrote in the early 1960's that "The 'control of nature' is a phrase conceived in arrogance, born of the Neanderthal age of biology and philosophy, when it was supposed that nature exists for the convenience of man."

Arguing that we were poisoning the entire planet with pesticides in our efforts to 'control' insects rather than learning to live along side them, Carson urged the world to halt "the chemical barrage [which] has been hurled against the fabric of life" (Carson 1964, 261).

In fact some sort of ecology of interconnection is a common theme articulated in feminist conceptions of knowledge (Daly 1978; Rich 1979; Griffin 1980; Merchant 1980; Lorde 1984; Keller 1985; Bleier 1986; Belenky et al 1986). In the work of both McClintock and Carson, this epistemology of interconnection is expressed through their careful attention to the dynamics of living systems as pieces of a larger and more awesome natural world which is constantly responding to, and responsive to, itself. As Haunani-Kay Trask has found in her analysis of other work by feminist writers, "their work reverberates with two themes: love (nurturance, care, need, sensitivity, relationship) and power (freedom, expression, creativity, generation, transformation)." These themes are what she had identified as "twin manifestations of the 'life force' " which she names "the feminist Eros" (Trask 1986, 86). We are now in a position to formulate an hypothesis: the hypothesis is that this "feminist Eros" will be an identifying landmark in the epistemology of gynocentric science. Yet while the examination of the work of women who have been recognized as scientists is exciting, it is also unsatisfying; we see only a small fraction of the work that women have done in investigating the world around them from their own perspectives. Following our hypothesis, and Keller's suggestion that a feminist conception of the erotic might yield a fundamentally different conception of science than the one that Plato bequeathed to us, it seems reasonable to suspect that gynocentric science in its natural habitat might already exist, looking somewhat different from androcentric science because of the different conception of the nature and position of the erotic with respect to epistemology (Keller 1985). It is this hitherto unrecognized science that I would like to begin uncovering here.

In searching through women's activities outside of those that have been formally bestowed the label of 'Science,' I have come to suspect that gynocentric science often has been called 'art,' as in the *art* of midwifery, or the *art* of cooking, or the *art* of homemaking. Had these 'arts' been androcentric activities, I have no doubt that they would have been called, respectively obstetrical *science,* food *science,* and family *social science.* Indeed as men have taken an interest in these subjects they have been renamed sciences—and, more importantly, they have been reconceived in the androcentric model of science. There is no question that all of these activities as defined and practiced by women have had important aesthetic, affective, social and erotic dimensions which androcentric science does not acknowledge as Scientific. But that is exactly what our hypothesis predicts for feminist science: that it will be less isolated from other aspects of our lives, less fetishized about individualism, more holistic, more nurturing, more concerned with relations than with objects, perhaps more dialectic, because of the nature and position of the erotic with respect to its epistemology. It might be that the presence of these aesthetic or affective components in gynocentric science underlies some of the reasons that the masculist guardians of Science have not recognized these activities as science, at least not as they were conceived and practiced by women.

GYNOCENTRIC SCIENCE IN ITS NATURAL HABITAT

In particular, I would like to suggest that we reexamine midwifery as a paradigm example of gynocentric science. This idea is not originally mine; in 1973 Ehrenreich and English suggested that the "magic" of the 16th Century European witch-healer and midwife "was the science of her time" (Ehrenreich and English 1973, 14). However by no means do I believe that midwifery is the only possible example of gynocentric science. For example, there are good reasons to believe that women's knowledge of food and nutrition historically had to include some fairly sophisticated knowledge of botany and ecosystems, as well as of human nutritional needs. This knowledge undoubtedly was accumulated as women worked in their capacities as the food, nutrition and health experts in a culture that had no pesticides, supermarkets or agricultural extension agents. Food production and preparation were largely women's provinces—and far from being mere social pastimes, they were indispensable to the sustenance of life. Without adequate knowledge of botany and of ecosystems, the life-sustaining gardens maintained primarily by women could have fallen victim to parasites, diseases and the depletion of soil nutrients. The food cultivated for humans would have been eaten by rabbits and deer and the genetic pool of the seed stock could have deteriorated through inbreeding; the concentrated plant populations in women's gardens would have been vulnerable to destruction by parasites and disease. Knowledge about the differences between edible and inedible plants, knowledge about the prevention of spoilage and food poisoning, knowledge about companion planting and crop rotation undoubtedly would have been part of the gynocentric science. One wonders whether food science would have gained the status of a science much sooner (and in a much different way) if women had been defining the sciences.

Pharmacology is another area in which there probably was a substantial tradition of gynocentric science. In his often quoted and equally often ignored remark, Paracelsus—the 'father' of modern pharmacology—attributed his entire knowledge of pharmacology to the wise women of his community. Paracelsus is an enigma in the history of science; modern androcentric historians of science don't know what to make of his mysticism and his holistic approach to the natural world. In the typically androcentric reconstruction of Paracelsus' work, his "Scientific" writings are unbundled from his "Unscientific" writings and packaged separately as Science, though we know that Paracelsus himself objected to this abuse of his own work. Given that he attributed his knowledge of pharmacology to a gynocentric tradition, it might be useful to us to reexamine *all* of what he did say, including that which is often omitted for being "unscientific," perhaps with the idea in mind of gaining clues to what gynocentric science looked like in his time.

But I doubt if gynocentric science is all from eras of long ago. In the twentieth century, for example, the androcentric social sciences have been granted the status of sciences, but the wisdom shared between and among women about the social fabric of their communities is still sneeringly labeled 'gossip.' Recently, however Belenky *et al* (1986) have suggested that gossip is a paradigm example of what they call "connected knowledge," a way of knowing that their research has found to be highly developed in some women. It would be interesting to investigate the idea that women's traditional vehicles of gossip—garden clubs, sewing circles, coffee klatches, baby showers and backyard fence discussions—actually are part of a gynocentric social science tradition which is oral and dialectic in nature. Another gynocentric field which has emerged in the last century is the field of home economics. Although this field has suffered from the attempts of androcentric educational administrators to turn its academic niche into vocational classes in cooking and sewing for future housewives, home economists undoubtedly have been concerned with home-based *economics,* a branch of economics that studies labor and production in the home. Additionally, there has been a very definite resurgence of gynocentric midwifery in the United States arising out of the women's health movement in the early 1970's. Often working outside of, or in opposition to, the law, lay midwives have organized schools, held conferences, published books and practiced midwifery in a way quite reminiscent of the midwives of the 16th and 17th centuries—who carried out their work in spite of the threat of the accusation of witchcraft for doing so (Ehrenreich and English 1973; Lang 1972; Arms 1975; Rich 1976; Gaskin 1978).

Women's Knowledge And Gynocentric Science

If I am correct that gynocentric science has existed all along, then there ought to be reasons that it has remained hidden to us, and I think that there are:

First, there is our training, which teaches us that 'science' is work that is done by people who have been awarded the title of Scientist, either by an institution from which they received a degree, or by history. Neither history nor degree-awarding institutions have been willing to award the title of Scientist to those who do not practice traditionally androcentric Western science, so it is to be expected that anyone who is practicing a different sort of science will not be included on the lists of certified Scientists. If we suspect that gynocentric science will be different in kind from androcentric science, even based in a fundamentally different epistemology, then it should be no surprise that gynocentric science would not be recognized as Science, and that gynocentric scientists would not have been labeled Scientists either by institutions or by history.

A second reason that it is easy to overlook gynocentric science is that the work of women has always been invisible in the recorded histories of androcentric Western culture. We are all familiar with the phenomenon of considering women's work to be non-work, as in "She doesn't work; she's a homemaker." Occasionally women scientists such as Marie Curie do make it into the recorded histories of science, but that is because Curie's work was spectacular and individualistic in nature, fitting it well within the model of what men think scientists do. If we are able to make out a case for two distinctly different scientific traditions, an androcentric tradition and a gynocentric tradition, we must not fall into the trap of believing that the practitioners within each of these traditions have been strictly divided along gender lines. Curie's work, though the work of a woman, was part of the androcentric tradition; the work of McClintock, Carson and Paracelsus probably was partially within each of the two traditions, as all three of these scientists seem to have been "claimed" to one extent or another as practitioners of both. Norman Casserley, a man prosecuted by the state of California for practicing midwifery in the early 1970's, probably was working within the gynocentric tradition (Arms 1975). The factors that identify a scientist as a member of one or the other tradition would be based in epistemological basis for inquiry, methodology, problem selection and scientific community—not in gender or sex. But the point is the same: not only women's activities, but all gynocentric activities, have been ignored by the androcentric recorders of history.

The third reason I would like to suggest is that throughout history the gynocentric sciences have been conducted primarily as oral rather than as written traditions. The accumulated wisdom and knowledge of midwives, for example, has been transmitted orally through personal contact and experiential apprenticeships. Midwives have not had professional journals in which they published their findings; they have not had heroes or methodological theoreticians who wrote treatises, nor have they had professional associations which held conferences and published their proceedings, and so on. Since Western androcentric science places such a premium on the written transmission of "results," any activity that has not included a large written component automatically has been excluded from consideration as a science. This undoubtedly is not unrelated to the fact that until very recently in Western history, literacy was much more available to White middle and upper class city-dwelling men than it was to women, the poor, Blacks, Native American tribes, rural families, and so on. Contrary to the usual assumption, which is that illiterate people don't do any science, I would like to propose that they probably do, but that their scientific traditions are oral and dialectic in nature rather than written.

A fourth reason, which is related to the third, is that women's knowledge, and its certification and transmission, may not have been organized hierarchically. The hierarchical organization of androcentric science was exquisitely described by Kuhn (1970) in his description of the socialization of young scientists into a scientific paradigm, and the rise and decline of particular paradigms. What he failed to realize, though Feyerabend (1975) did not, is that the hierarchical organization of both scientific knowledge and scientists is not a *necessary* feature of science, but rather simply a feature of Western science to which we have grown accustomed. While Feyerabend's epistemological anarchy might not provide a good description of the organization of gynocentric knowledge, it does provide the imagination with an alternative to the hierarchical organization of Western scientific knowledge.

A fifth reason, readily visible in the cases of midwifery and 'gossip,' is that there has been a concerted effort to suppress and discredit these bodies of gynocentric knowledge as being erroneous, based in superstition, and connected with harm or evil. More chillingly, history has recorded—although it attempts to forget—a violent campaign of torture and murder in the European witchburnings which was both implicitly and explicitly directed at eradicating various gynocentric traditions (Ehrenreich & English 1973; Daly 1978; Edwards & Waldorf 1984). The influences of force and violence in establishing the Western androcentric scientific tradition as the dominant one should not be underestimated.

Science And Bodily Functions

A look at the gynocentric approach and the androcentric approach to childbirth may provide a case study of one of the longest running disputes between—as Kuhn puts it—competing paradigms. The incommensurability between a gynocentric and an androcentric point of view is clearly visible in the differences between midwives' and scientific doctors' conceptions of childbirth. While medical science views childbirth as an abnormal state of health that has the potential to develop into a serious emergency, midwives have taken a much more holistic view of childbearing than has medical science; childbirth has been viewed as a normal physiological function during which a woman has an increased need for community support. This is not to say that midwives hold the naive view that life- or health-threatening emergencies never arise in the course of childbirth. Occasionally they do. But people also occasionally choke on food, drown in bathtubs, suffer strokes while playing golf and die of heart attacks during sex.

Science Fiction: Consider the following imaginary scenario: Suppose that, recognizing the many possible dangers associated with eating mishaps, the society decided to get much more scientific about the whole process. Federal funds were allocated for setting up elaborate hospital dining halls, and everyone who ate or served food was first urged, then required, to do so under proper medical supervision, "just in case." After all, one could never predict when people might choke on, or have a sudden allergic reaction to, their food. Additionally, experts in the eating sciences had become increasingly concerned about the lack of sterile conditions under which food was typically prepared in the home. Science had already well documented the large numbers of bacterial organisms found in virtually every home kitchen, and its increasing knowledge of the role of bacteria in disease made it obvious that untrained cooks, usually women, in bacteria-laden kitchens could no longer be trusted with the important responsibility of feeding the general population. There was particular concern about the health of children fed in the home. Young children could unknowingly be fed foods to which they might have a violent allergic reaction, and food scientists were alarmed about the numbers of children who were being exposed to the possibility of hives, asthma, lactose intolerance and digestive disturbances during home feeding. It was also suspected that Sudden Infant Death Syndrome and perhaps even some learning disabilities might be linked to unscientific feeding practices. But the dangers of home feeding were not confined to young children. It had long been recognized that eating practices were large factors in adult onset diabetes, heart disease, obesity, anorexia nervosa, diseases of the digestive system, cancer and possibly even drug abuse. At the hospital, each person could be carefully monitored for weight, calorie consumption, vitamin intake, and the percentage of fat and fiber in the diet. Special health problems could be discovered immediately, and each patient's blood and urine could be monitored regularly by the physicians who attended their eating sessions. High risk patients also might have their stomach secretions and peristaltic contractions electronically monitored during each meal at a central nursing station, and a computer controlled alarm would sound if any diner's digestive readings were outside the normal range for the type of meal that was being consumed. Some experts were even starting to suggest that all diners be electronically monitored while they were eating. If there was any indication of a developing eating problem, the patient would probably be admitted to the hospital for intravenous feeding until the problem had cleared up.

Science Reality: This is what androcentric science has done to childbirth.

Human Birth: A Review

As mammals, humans have always given birth to live young. Human pregnancy and childbirth, like eating, always have been biologically successful; that is, they have not been so hazardous or dangerous as to threaten the biological success of the human species. Unlike disease, which threatens the life or the fitness of individuals who suffer from it, pregnancy, childbirth and lactation normally do not threaten the fitness of either mother or child. If pregnancy, childbirth and lactation usually resulted in dead or unfit offspring, or in dead or incapacitated mothers, our species would not have survived. The fact is that the human species not only survives, it thrives with respect to reproduction, particularly when adequate nutrition, shelter and freedom from disease are available to mothers and their offspring. Studies have shown repeatedly that the vast majority of all human pregnancies would end with the birth of a healthy infant to a healthy woman even if she received no prenatal care or advice of any kind from any source (Lang 1972; Guttmacher 1973; Arms 1975; Rich 1976; Oxorn & Foote 1980). With social support, proper nutrition, adequate attention to physical fitness, and the elimination of habits detrimental to health such as smoking, drinking, and drug abuse, some studies have suggested that this figure may approach 95–98% or higher (Lang 1972; Guttmacher 1973; Arms 1975; Gaskin 1978; Edwards & Waldorf 1984). There is documented evidence, for example, that maternal mortality rates were approximately 0.4% in midwife-attended births in the American colonies during the eighteenth century—well before the introduction of modern antibiotics which could cure life-threatening infections (Wertz and Wertz 1979). Midwife-attended home births in Leslie County, KY, one of the poorest rural areas in Appalachia, with the highest birth rate in the country, had a maternal mortality rate of 0.091% from 1925–1955, compared to the national average of 0.34% for white women during the same time period (Arms 1975). This is consistent with the expected levels of health following birth in other species of mammals with typically single gestations. Often touted pseudo-explanations of Western women's difficulties in childbirth simply do not hold up under examination. Evolutionary biology, for example, does not support the idea that Western women could have "evolved" into a species with pelvic structures unsuited for childbirth in the dozen or so generations since androcentric doctors became interested in childbirth. It is possible that women's general levels of physical conditioning and nutrition could have deteriorated over this time period in such a way that the typical Western woman is in poor physical health throughout her childbearing years, but this is entirely environmental, not genetic.

It is also well documented that many of Western women's difficulties in giving birth are the result of the conditions imposed upon them by Western hospitals (Oxorn & Foote 1980; Arms 1975; Mendelsohn 1982). A prime example of this is the condition known as supine hypotensive syndrome. Oxorn & Foote describe the condition this way:

> The clinical picture is one of hypotension when, in the late stages of pregnancy, the woman lies on her back. . . . Other symptoms include nausea, shortness of breath, faintness, pallor, tachycardia, and increased femoral venous pressure. . . . Reduced perfusion of the uterus and placenta leads to fetal hypoxia and changes in the fetal heart rate. (Oxorn & Foote 1980, 115)

This condition may occur to some extent or another in virtually all women who give birth in hospitals, or approximately 97.4% of all births in the United States. This is because hospital births are conducted with the laboring woman ("in the late stages of pregnancy") lying on her back in a bed or on a delivery table. The most common position, known as the lithotomy position, has the woman flat on her back with her legs up in the air in stirrups. Oxorn & Foote list as advantages of this position: more complete asepsis, easier for hospital personnel to monitor the fetal heartbeat without asking the woman to change position, easier for hospital personnel to administer three different kinds of drugs, easier for the doctor to see the birth, good position for the use of forceps and for performing the surgical procedure of episiotomy. As disadvantages they list: risk of supine hypotensive syndrome, sacroiliac or lumbosacral strain, possible thrombosis in the veins of the legs, possible nerve damage, and the danger of aspiration of vomitus (Oxorn & Foote 1980, 114). In contrast, they list the following advantages and disadvantages for women's and midwives' more traditionally preferred squatting position, which is almost never allowed by hospital regulations: Advantages: enlarges the pelvic outlet, enables laboring woman to use her expulsive forces to the greatest advantage, eliminates risk of supine hypotensive syndrome; Disadvantages: difficult "for the accoucher to control the birth and to manage complications," impossible to administer certain types of drugs (Oxorn & Foote 1980, 113).

A feature of the lithotomy position not mentioned by Oxorn & Foote, perhaps because they couldn't decide whether it was an advantage or a disadvantage, is that it increases the amount of symptoms which are taken as indications for performing a cesarean section. The primary fetal indications for cesarean surgery are fetal hypoxia and changes in the fetal heart rate (Oxorn & Foote 1980, 667). These are exactly the fetal conditions that are part of the clinical picture of supine hypotensive syndrome. One would think that since Oxorn & Foote report "the risk of maternal death associated with cesarean section to be 26 times greater than with vaginal delivery" (Oxorn & Foote 1980, 675) the avoidance of cesarean section would be an advantage.

However they continually reassure the reader that cesarean section is a fine thing, a tribute to progressive technology. In fact, they advise against the cesarean operation only "when the fetus is dead or in such bad condition that it is unlikely to survive" because "[t]here is no point in submitting the patient to a needless serious operation" (Oxorn & Foote 1980, 669). One cannot help but suspect that physicians who have undergone nearly a decade of training in techniques of medical intervention find it boring and wasteful of their expertise *not* to engage in such interventions whenever possible.

Obviously in this case, when the laboring woman's physiological needs conflict with the convenience or the need for "control" on the part of the hospital staff, it is the woman's physiological needs that are sacrificed. One can hear echoes of Rachel Carson's protest about the "control of nature" and the dangerous effects of that approach in the critiques made by gynocentric investigators about androcentric conceptions of childbirth. Other factors may color androcentric science's research results as well. As Dr. Robert Mendelsohn put it in *Mal(e)practice,* "Doctors know that they can't afford to allow their patients to perceive childbirth as the normal, typically uncomplicated process that it really is. If they did, most women wouldn't need obstetricians" (Mendelsohn 1982, 130). One can almost imagine the same words being written by a dissident physician with respect to eating, in the years following the imaginary scenario in which androcentric science takes control of eating practices. These things are important to keep in mind as we compare gynocentric midwifery with androcentric obstetrical science.

Midwifery and Obstetrics as Competing Scientific Paradigms

Obstetrical science, for the most part, has adopted the view that midwifery is an incomplete, underdeveloped, less successful, and less scientific approach to the same scientific problems that it is attempting to solve. One cannot help but note the similarities between this view of midwifery and the now discredited psychoanalytic and philosophical theories that make out women to be incomplete, underdeveloped, less successful and less rational versions of men. A less biased description might be that midwifery and obstetrical science represent competing scientific paradigms which, like all competing paradigms according to Kuhn (1970), disagree not only about the list of problems to be resolved, but also about the theories, methodologies, and criteria for success that will be used to assess the results achieved.

If gynocentric science is, as I've suggested earlier, "less isolated from other aspects of our lives, less fetishized about individualism, more holistic, more nurturing, more concerned with relations than with objects, perhaps more dialectic, because of the nature and position of the erotic with respect to its epistomelogy," then these aspects of it should be evident in the work of gynocentric scientists practicing within their paradigm. On the other hand, the androcentric criteria for good science such as abstraction, reductionism, the determination to repress one's feelings to promote 'objectivity' cited by Namenwirth (1986) should be evident in the work of androcentric scientists practicing within their paradigm. Consider these two examples:

MIDWIFERY AS GYNOCENTRIC SCIENCE

. . . I want to stress the importance of good continuous prenatal care. Without the knowledge of excellent health the risks of home birth increase for both mother and child.

To have a healthy pregnancy and good childbirth, certain aspects of existence on the physical, mental, and spiritual plane must be observed and trained to be in harmony with the forces within you, i.e., that of the creation of life.

You should be able to listen closely to everything your body is telling you about what's happening within, how your body feels about what goes into it, what comes out, and just how it feels organically.

As the pregnancy proceeds many things happen and a gradual process of training mind and body takes place.

Food may become a necessary discipline if a diet is not made normally of whole foods. Foods that have been flash grown, processed, refined, nutritionalized and put out as some predigested matter should be avoided.

. . . It's pretty easy to tell if you're doing the above correctly because you will be healthy organically and you can feel that through your entire body.

—*Raven Lang*
Midwife

OBSTETRICS AS ANDROCENTRIC SCIENCE

Once the patient has carefully selected her doctor, she should let him shoulder the full responsibility of her pregnancy and labor, with the comforting knowledge that, no matter what develops, he has had similar cases and her health will be safeguarded by this background of experience.

Most obstetricians prefer to see their patients early in pregnancy, two or three weeks after the first menstrual period is missed. Many women look forward to this first interview with unnecessary dread. Perhaps a friend with previous experience has told them that it is a most embarrassing examination, and such questions! The patient is apt to forget that the doctor has examined literally thousands of women, and in the course of this experience has learned to impersonalize his attitude toward his patients.

At the first visit the obstetrician examines the woman completely from top to toe. It is essential that he determine the exact physical condition

> of his patient so that he may judge her ability to withstand the strain of pregnancy and labor.
>
> —*Dr. Alan Guttmacher,*
> Professor Emeritus of the Department of Obstetrics and Gynecology at New York's Mount Sinai Medical School

If incommensurability is to be taken seriously, then we are faced with the ever present problem of theory choice. How are we to evaluate two competing scientific paradigms with respect to their successes at problem solving? "In the first place," wrote Kuhn "the proponents of competing paradigms will often disagree about the list of problems that any candidate for paradigm must resolve. Their standards or their definitions of science are not the same." Going on to explain linguistic incommensurability, he then comes to one of the most compelling observations in *The Structure of Scientific Revolutions:* "In a sense that I am unable to explicate further, the proponents of competing paradigms practice their trades in different worlds" (Kuhn 1970, 150). While the androcentric scientists scratch their heads in perplexed confusion about what Kuhn might possibly mean by this, many feminists smile with the pleasurable sensation of having encountered a clear articulation of the obvious. Many of us are well acquainted with the feeling of dividing our time and attention between two different worlds. Carol Gilligan (1982) has suggested ways in which this plays itself out in ethics; Belenky, Clinchy, Goldberger, and Tarule (1986) have recently done the same for epistemology. There is no reason for us to suspect that we can't do the same thing with respect to science. I don't for a minute want to claim to have demonstrated the existence of a gynocentric science here; what I have tried to do is to articulate an hypothesis that deserves further investigation. My hope is that some empirical investigation will yield fruitful results in this direction, and that these results—in the best of scientific traditions—can feed back into, and interact with, our growing body of theory about the nature of feminist science.

WHY CALL IT 'SCIENCE?'

Well, why not?

Women *are* trapped in an androcentric world, as Bleier suggests, one in which language and meaning have been constructed around androcentric goals and enterprises. We've had troubles with language all along. As Marilyn Frye has pointed out, the very terms we use embed in them the connections and distinctions that *men* want to see (Frye 1983, 161). If the term 'science' is to be construed only as a limited range of activities, conducted by properly certified people, under a limited range of circumstances, then perhaps the term 'gynocentric science' is as much a self-contradiction as the term 'military intelligence.' But one of the projects of American feminists has been to claim our right to participate in the making of meaning. We have struggled, for example, to be able to apply the term 'scholarship' to our work, even when much of that work didn't count as scholarship under the old androcentric language rules. For that matter, we've struggled for the right to apply the term 'work' to many of our activiites that were once considered not to be work. Feyerabend (1975) noticed, even without the benefit of a feminist perspective, that the distinction between science and non-science is political. And as Frye (1983, 105) pointed out, "*definition* is another face of power."

So maybe we are recreating the language a bit by calling midwifery or gossip or cooking 'gynocentric science.' But then, as members of the language-using community, we are entitled. The burden is not on us, but on those who object, to show that this is not a reasonable use of the term 'science.'

HESITATIONS

Even as I write this, I'm not completely convinced that tugging on the term 'science' to fit gynocentric activities under its umbrella is necessarily the right thing to do. Perhaps, as Marion Namenwirth (1986) suggests, "abstraction, reductionism, the determination to repress one's feelings to promote 'objectivity' have not the same priority to women as they do men." Perhaps, as she doesn't suggest, these qualities are already so tied to the term 'science' that we will choose to dissociate our work from the baggage of the term, and name our gynocentric work something else instead. I find that my own feelings about this waver. On the days when I'm hoping that feminism will make the world better for everybody, I want to tug at the meaning of science to get it to include gynocentric activities. On the days when I'm seeing science as the religion of advanced patriarchy, and philosophy as its theology, I want to withdraw from both entirely. But for those of us who, at least on some days, are struggling to find a feminist conception of science, I offer this suggestion: it's been around us all along. Our task now should be to research it; we need no longer merely fumble about for a theory of what it might be.

REFERENCES

Arms, Suzanne. 1975. *Immaculate deception: A new look at women and childbirth in America.* Boston: Houghton Mifflin.

Belenky, Mary *et al.* 1986. *Women's ways of knowing.* New York: Basic Books.

Birke, Linda. 1986. *Women, feminism and biology: The feminist challenge.* New York: Methuen Press.

Bleier, Ruth. 1984. *Science and gender: A critique of biology and its theories on women.* Elmsford, NY: Pergamon.

———, ed. 1986. *Feminist approaches to science.* Elmsford, NY: Pergamon.

Carson, Rachel. 1964. *Silent spring.* New York: Fawcett Crest.

Daly, Mary. 1978. *Gyn/ecology: The metaethics of radical feminism.* Boston: Beacon Press.

Edwards, Margot and Mary Waldorf. 1984. *Reclaiming birth.* Trumansburg, NY: Crossing Press.

Ehrenreich, Barbara and Deirdre English. 1973. *Witches, midwives, and nurses: A history of women healers.* Old Westbury, NY: The Feminist Press.

Fee, Elizabeth. 1983. Women's nature and scientific objectivity. In *Women's nature: Rationalizations of inequality,* ed. M. Lowe and R. Hubbard, 9–28. Elmsford, NY: Pergamon.

Feyerabend, Paul. 1975. *Against method.* London: Verso.

Frye, Marilyn. 1983. *The politics of reality: Essays in feminist theory.* Trumansburg, NY: Crossing Press.

Gaskin, Ina May. 1978. *Spiritual midwifery.* Summertown, TN: The Book Publishing Company.

Gilligan, Carol. 1982. *In a different voice.* Cambridge, MA: Harvard University Press.

Griffin, Susan. 1980. *Woman and nature: The roaring inside her.* New York: Harper and Row.

Guttmacher, Alan. 1973. *Pregnancy, birth and family planning.* New York: Viking Press.

Harding, Sandra. 1986. *The science question in feminism.* Ithaca, NY: Cornell University Press.

Keller, Evelyn Fox. 1983. *A feeling for the organism: The life and work of Barbara McClintock.* San Francisco: W. H. Freeman.

———. 1985. *Reflections on science and gender.* New Haven, CT: Yale University Press.

Kuhn, Thomas S. 1970. *The structure of scientific revolutions.* Chicago: University of Chicago Press.

Lang, Raven. 1972. *Birth book.* Ben Lomond, CA: Genesis Press.

Lorde, Audre. 1984. *Sister outsider.* Trumansburg, NY: Crossing Press.

Mendelsohn, Robert. 1982. *Mal(e) practice.* Chicago: Contemporary Books.

Merchant, Carolyn. 1980. *The death of nature: Women, ecology and the scientific revolution.* New York: Harper and Row.

Namenwirth, Marion. 1986. Science seen through a feminist prism. In *Feminist approaches to science,* ed. Ruth Bleier. Elmsford, NY: Pergamon.

Oxorn, Harry and William Foote. 1980. *Human labor & birth.* New York: Prentice-Hall.

Rich, Adrienne. 1976. *Of woman born: Motherhood as experience and institution.* New York: Norton.

———. 1979. *On lies, secrets, and silence: Selected prose—1966–78*. New York: Norton.

Rose, Hilary. 1986. Beyond masculinist realities: A feminist epistemology for the sciences. In *Feminist approaches to science*, ed. Ruth Bleier, 57–76. Elmsford, NY: Pergamon.

Stone, Merlin. 1978. *When God was a woman*. New York: Harcourt Brace Jovanovich.

Trask, Haunani-Kay. 1986. *Eros and power*. Philadelphia: University of Pennsylvania Press.

Wertz, Richard and Dorothy Wertz. 1979. *Lying-in: A history of childbirth in America*. New York: Shocken.

Justifying Feminist Social Science

LINDA ALCOFF

In this paper I set out the problem of feminist social science as the need to explain and justify its method of theory choice in relation to both its own theories and those of androcentric social science. In doing this, it needs to avoid both a positivism which denies the impact of values on scientific theory-choice and a radical relativism which undercuts the emancipatory potential of feminist research. From the relevant literature I offer two possible solutions: the Holistic and the Constructivist models of theory-choice. I then rate these models according to what extent they solve the problem of feminist social science. I argue that the principal distinction between these models is in their contrasting conceptions of truth. Solving the problem of feminist social science will require understanding that what is at stake in the debate is our conception of truth. This understanding will serve to clarify, though not resolve, the various approaches to and disagreements over methodologies and explanations in feminist social science.

THE PROBLEM

The very existence of a category called feminist social science creates a philosophical problem. For the category implies a relationship between feminism and social science that is not merely coincidence, as if it were a matter of chance that these social scientists happened also to be feminists. In fact, the relationship of these social scientists to their feminism is not one that merely affects their choice or priority of research subjects. The relationship is actually much stronger—one could say that the feminism of these social scientists enters into not only the context of discovery but also the context of justification in their work. For what we find is feminist social scientists developing and defending what might be called feminist-consistent theories, that is, theories that are broadly speaking consistent with and supportive of the general aims, aspirations, assumptions and even political programs of feminists. In other words, theories in feminist social science are not simply theories about women, but theories of a particular type about women (i.e., exhibiting consistently emancipatory implications for women's position in society). I do not think it is necessary to enumerate here all the work done

Hypatia vol. 2, no. 3 (Fall 1987). © by Linda Alcoff.

by such scientists as Louise Lamphere, Michelle Rosaldo, Sherrie Ortner, Carol Gilligan, Dorothy Smith, Elizabeth Stanley, Sue Wise, Nancy Chodorow, Harriet Whitehead, Karen Sachs, Naomi Weisstein and many others to demonstrate the persuasiveness of this point. Although there are disagreements, often heated, within and among this group of social scientists, there is a fairly clear family resemblance in these works which sets them apart from other work in the social sciences.

What we must ask, however, is: what's going on here? It would be easy at this point to find an explanation of how feminist social scientists develop feminist-consistent theories by turning to a sociologist of knowledge, but this sort of explanation is not satisfactory to me, at least not yet. The problem as I see it is not so much to wonder how it is feminists develop such theories, but to wonder what method of theory-choice is being used here that furnishes such results. Specifically, as Elizabeth Potter has pointed out, feminist social scientists need to answer the following two questions:

> On what grounds do we challenge non-feminist and particularly androcentric assumptions and theories? and: On what grounds do we argue for the superiority of feminist assumptions and theories? (Potter 1984, 1)

These questions in essence boil down to one: what is the method of theory-choice in feminist social science?

Moreover, we need both an explanation of and a justification for the method of theory-choice in feminist social science, an explanation that can account for the relationship between the feminism and the social science in some other way than attributing it to coincidence, and a justification that shows why our respect for this general body of research is not ill-founded. Of course, we need not nor do we want a justification of each and every claim or theory advanced by a feminist social scientist, but this just complicates our endeavor. For we need an explanation and justification of feminist social science as a research program while retaining our ability to evaluate better and worse theories advanced within this program according to some significant criterion.

The care we take in answering these questions is prompted by a background concern that has been articulated best perhaps by Evelyn Fox Keller:

> Joining feminist thought to other social studies of science brings the promise of radically new insights, but it also adds to the existing intellectual danger a political threat. The intellectual danger resides in viewing science as pure social product; science then dissolves into ideology and objectivity loses all intrinsic meaning. In the resulting cultural relativism, and emancipatory function of modern science is negated and the arbitration of truth recedes into the political domain. (1982, 117)

It certainly does seem as if much, if not all, of the motivation behind the recent upsurge in feminist social science has been precisely because of what is perceived as its "emancipatory function," a view prompted by the truism that the first step to enable a man to the achieving of great designs is to be persuaded that by endeavor he is able to achieve it. Though in this case it is a woman, or all of womankind and even mankind, that is to be persuaded, through the scientific study of the mind, brain, behavior and social structures of human beings, that woman's liberation from her universal position of subordination is at least possible. No amount of ethical reasoning will help if this first premise is proved false. And so the impetus for a large amount of empirical research appears to be primarily and unmistakably political.

Identifying the political impetus for feminist social science does not yet entail "receding into the political domain" that Keller worries about, however. Having political motivations will not necessarily turn science into ideology until and unless we replace scientific theory-choice with political argumentation.[1] And yet the coincidence of feminist-consistent theories being churned out by feminist social scientists might lead one to believe that that replacement has already occurred. Again, what we need is an account of the theory-choice operative in feminist social science that will provide plausible and coherent answers to our two initial questions while waylaying the concerns expressed by Keller. Our project, moreover, must be consistently self-critical even while it is externally directed. That is, we must guard against making uncritical assumptions concerning which accounts of theory-choice are "plausible and coherent." While we should not want to accept a model of theory-choice simply and entirely because it satisfactorily explains the existence of feminist social science (i.e., without disqualifying it as a science), neither should we uncritically dismiss feminist social science if it cannot be accounted for by any pre-existing mainstream model of theory-choice. Our critical emphasis needs to be divided equally between the objective of our problematic and the methodology we choose to pursue. In analyzing feminist social science we may be likely to run head on into an analysis of models of theory-choice generally, both those models accepted within the mainstream debates and those models excluded. To presume an arbitrary limit on the potential of feminist social science for radically challenging dominant views in the sciences is to constrain our objective in such a way that it might be finished before it is begun.

With these cautions in mind, let us proceed.

The Possible Solutions

Now it may seem as if there is already available a straightforward and entirely acceptable justification and explanation of feminist theory-choice. Previous androcentric theories in social science are explained as simply the product of masculine bias, or masculine desire to protect the status quo, and since feminist social scientists have no such interest in maintaining the status quo they are free to see and report the data clearly and without danger of bias. Many feminist social scientists make this sort of a claim, from Margaret Eichler (1980, 51) and Evelyn Reed (1978, 7) to Nancy Schrom Dye (1979, 22–28) and Ruth Hubbard (1983, 45–71).[2] Such an explanation relies heavily on the obvious empirical insufficiency of many theories rejected by feminists, from penis envy to sociobiology, and the conclusive empirical substantiation of (some) theories within feminist social science. The feminist project, under this sort of an explanation, becomes the process of identifying and removing the elements and effects of bias within mainstream social science, in effect purifying social science of its non-objective elements.

This approach of course assumes generally speaking a positivist conception of theory-choice in which empirical determination plays a decisive role and processes of theory-development and theory-justification are *in practice* capable of being value-neutral.[3] It posits the problem in past social science as the wrongful interdiction of values into the realm of science, so the solution is simply to remove those values and return science to its proper procedures. If this explanation is correct, however, we must assume and eventually be able to prove that the pure objectivist (and available) methodology of theory-choice was uniformly rejected, consciously or unconsciously, by all previous social scientists who produced erroneous theories about women—that none of them were careful, intellectually honest, socially progressive in relation to gender roles, or simply good scientists. Surely this claim will be difficult if not impossible to prove.[4]

More importantly, the positivist conception of theory-choice as essentially empirically determined or at least value-neutral has come under enough severe and sustained attacks in recent times to make it wholly unacceptable in my view. Specifically, as Sandra Harding (1976, ix–xxi) has argued, Pierre Duhem's (1962, especially chapter six) compelling arguments that all propositions in physics are empirically underdetermined and Thomas Kuhn's (1970) description of the history of science as paradigm-guided have seriously eroded many philosopher's of science commitment to the positivist model. The fact that the evidence for these claims comes from physics and not the social sciences makes it even less likely that social science can claim adherence to an ideal of methodological objectivity that even physics must forego. Moreover, the mounting evidence of value-laden theory-choice throughout the sciences seems to substantiate the view that there is no value-free methodology, or at least not one that human scientists can realistically employ *in practice*. Hilary Putnam makes this argument succinctly as follows:

> . . . the concern of exact science is not just to discover statements which are true, or even statements which are true and universal in form ('laws'),

> but to find statements that are true and *relevant*. And the notion of *relevance* brings with it a wide set of interests and values. (Putnam 1971, 137)

To the extent that all theories are empirically underdetermined, we must look for more than lack of evidence to explain androcentric theory-choice. Our conception of feminist theory-choice, likewise, cannot rely on our ability to prove that feminist social scientists have been more objective (in the positivist sense) than their predecessors in the field.

Another popular and apparently straightforward explanation of feminist social science involves the claim counter to positivism that all scientific theory-choice is value-laden but that the values involved in feminist theory-choice are in a number of epistemically interesting ways better than the values involved in previous non-feminist or androcentric theory-choice. Thus it is argued sometimes that the bias feminists use is a human bias rather than a male bias, and so more complete and comprehensive. The British sociologists Liz Stanley and Sue Wise exemplify this sort of view when they claim:

> feminism . . . demonstrates that the social sciences are sexist, biased, and rotten with patriarchal values. However, feminist social science can be truly scientific in its approach. Having eradicated sexism, we can see and research the world as it truly is. Feminism encapsulates a distinctive value position, but these are truly *human* values . . . (Stanley and Wise 1983, 12)

Notice that Stanley and Wise have no trouble in juxtaposing the claims that feminist social science can be truly scientific and that it involves a distinctive value position.

Obviously, if we reject the possibility of a value-neutral model of theory-choice to explain feminist social science as I have done then we are forced to adopt some form of a value laden model of theory-choice, as Stanley and Wise do. But, if we share Keller's concerns we must admit that there is a real danger of radical relativism here.[5] Uncareful articulations of value-laden theory-choice could lead to the view that the differences between feminist and androcentric social scientific theories are *merely* political differences, which do not render the theories true and false but only different. In other words, it might lead to a conception of social scientific theories as having an empirical content and a value content, autonomous from each other in such a way that two conflicting theories may share their empirical content but contradict each other's value content. In such a scenario, the debate over theory-choice would necessarily recede to the political and/or ethical domain entirely.

There is a strong sense that such a situation would decisively undercut the persuasiveness of feminist social science theories, and certainly their emancipatory potential. Feminist social science would become, in fact, superfluous to other forms of feminist struggle, unable to provide any independent corroboration or defense of feminist claims. Moreover, if the "hard" empirical evidence is inconclusive or undecidable enough to imply or justify both antifeminist and feminist conclusions, the political struggle of feminists would undoubtedly become that much more difficult.

What we need, then, is a careful working out of a model of theory-choice which (a) avoids implicating feminist conclusions in a radical relativism which involves a total replacement of empirical evidence with political debate, and (b) explains and justifies in a plausible and coherent manner the history of social science in which androcentric theories were developed and defended mostly by males, and feminist-consistent theories more recently have been developed and defended mostly by females. As if this were not enough, we also need a model that allows us to evaluate theories in feminist social science by more than the criterion of being feminist-consistent. As stated, in my view such a model can only be found among attempts which integrate values into the context of justification, which show values to be epistemically significant and not merely unavoidable distorting influences inevitably interfering in the process of theory justification. Some value-laden models of theory-choice appear to echo the positivist view that values are inevitably distorting and epistemically irrelevant; what is argued, then, is simply that we cannot have a pure value-free theory-choice so it is in vain to construct philosophies of science based on one. Some of Kuhn's writings seem to harbor such a position. A much stronger position would be to argue that values are legitimate, and not merely unavoidable, components in theory-choice. The danger here again, however, is the possibility of slipping into a radical or absolute relativism which undercuts both our persuasiveness and our coherence.[6]

TWO MODELS OF THEORY-CHOICE

From my readings in philosophy and feminist theory I have found what can be classified as two general types of solutions to the problem of feminist social science. Whether either is truly a solution is still open to question, but the following comparative explication of them should advance our ability to answer that question.

The first is the Holistic Model of theory-choice. This model borrows heavily from the writings of C. S. Peirce (1940), W. V. O. Quine (1963, 20–46), and Mary Hesse (1976, 184–204; 1980), and is exemplified in the feminist theoretical work of Sandra Harding (1977, 351–366; 1983; 1986), Elizabeth Potter (1984), and Evelyn Fox Keller (1982, 1985).[7] I believe it can be demonstrated that this model is also working implicitly in the theoretical work of many other feminist theorists, but such a demonstration cannot be undertaken here.

The hallmark of the Holistic Model is that it uses a coherence criterion for theory-choice within a web of belief. Quine explains the way this model works in the following way:

> Total science is like a field of force whose boundary conditions are experience. A conflict with experience at the periphery occasions readjustments in the interior of the field . . . But the total field is so undetermined by its boundary conditions, experience, that there is much latitude of choice as to what statement to reevaluate in the light of any single contrary experience. No particular experiences are linked with any particular statements in the interior of the field, excepting indirectly through consideration of equilibrium affecting the field as a whole. (Quine 1963, 42–43)

The impact of this model on our conception of theories as part empirical content/part values can be seen from the following passage.

> If this view is right, it is misleading to speak of the empirical content of an individual statement . . . Any statement can be held true come what may, if we make drastic enough adjustments elsewhere in the system. Even a statement very close to the periphery can be held true in the face of recalcitrant experience by pleading hallucination or by amending certain statements of the kind called logical laws. (Quine 1963, 43)

Thus, the values and empirical content of any statement cannot be discretely differentiated. Coherence is the criterion we use in theory-choice, but we cannot hierarchize that which a theory must be coherent to other than to distinguish the more central and the more peripheral parts of our web of belief. The center of the web will contain those statements held more in common by more theories, those statements which will require the greatest amount of adjustments throughout the web if they are to be adjusted themselves. It may be that claims about women's subordination have held just such a central place in our web of belief, which would explain why the elimination of these claims is now motivating a significant number of feminist scholars to assert that we need to make radical and far-ranging readjustments throughout the entire web. Theory-choice thus involves, on this model, not the mere linking up of individual empirical claims to similarly limited though more general empirical claims (as on the positivist model), but involves to various degrees of significance the entire web. To explain feminist social science we can plug in a concept such as Kuhn's paradigm, and then be able to account for feminist social science without making either androcentric or feminist social scientists simply irrational, biased or bad scientists.

The Holistic Model claims to avoid radical or absolute relativism because, first, it admits that experiential anomalies must be accounted for *somehow*, even if only by hallucination, thus eliminating the possibility that we are totally unconstrained in what we choose to believe. Moreover, a separate argument could be advanced to indicate that it is at least highly likely that some of our beliefs within this web are true, even though we cannot at any time designate which those are. This model also avoids replacing the scientific for the political domain by squarely implanting political beliefs within the process of scientific (or epistemic) theory-choice.

We will return to explore the Holistic Model further in a moment, but now we will introduce our second model of theory-choice, and then proceed to a comparison. I call the second model the Constructivist Model of theory-choice, and it relies most heavily on the writings of Michel Foucault (1972, 1980) and Hans-Georg Gadamer (1976, 1984). (As heterogeneous as these authors are on some topics, such as the epistemic significance of the subject, in my view their theories intersect at a clear and interesting point in the following model.) Feminist theorists such as Marcia Westkott (1979, 422–430) and Dorothy Smith (1977, 15–22), and to a somewhat less clear extent Nancy Hartsock (1983) and Teresa de Lauretis (1984), exemplify the Constructivist Model in their work.

The Constructivist Model again relies on coherence as the criterion of theory-choice but adds a new focus on a discourse which sets out the prior background conditions necessary before a statement can even be considered. Discourse represents, for the hermeneuticist, the meta-layer of "common meanings" inaccessible and yet necessary for justification.[8] Similarly for Foucault, "before a proposition can be pronounced true or false it must be, as Monsieur Canguilhem might say, 'within the true' " (1972, 224). Justification by coherence works at this meta-level but only by implicit and imperceptible means. Thus, pronouncements of women's equality, on this view, were not even statable in the past. However, discourse, for Foucault at least, must not be conceptualized as always acting as a constraint on knowledge-claims, as an outer limit of the statable, but rather as a producer of knowledge, in a somewhat similar way that the primary role of paradigms in Kuhn's sense is to provide problems for the scientists to solve, rather than simply narrowing the sphere in which it is possible to work. In saying that discourse produces and not merely constrains knowledge the point Foucault seems to want to make is that power and/or politics, as central features of the discourse, are likewise not simply distorting factors, ever present obstacles to our pursuit of the truth, but in some sense *constituents of truth*. Thus the Constructivist Model clearly does not relegate values to the role of epistemic distortions.

On this model, feminist social science might be seen as a discourse. Similarly to the Holistic Model, the Constructivist Model explains feminist social science within the coherence of a discourse, which looks to be similar again to (though broader than) a Kuhnian paradigm. The Constructivist Model also furnishes a way of conceptualizing the integration of values into the epistemic realm and thus avoiding a simple replacement of science with politics. However, it may seem as if the two models are thus far not significantly distinct in terms of their explanation of theory-choice. Moreover, the Constructivist Model has not demonstrated how it avoids the problem of radical relativism at all, whereas the way in which the Holistic Model avoids this problem may be as yet unclear. What I find to be the important distinction between these two models is their contrasting conceptions of *truth* (which necessarily involves contrasting ontologies). The explication of these two conceptions of truth should advance our understanding of these two models of theory-choice, their distinctiveness from one another, and the ways in which they seek to successfully meet the various criteria for a model of theory-choice expressed above.

TWO CONCEPTIONS OF TRUTH

The Holistic Model operates with a correspondence theory of truth even while it uses coherence as a criterion of truth. Mary Hesse writes of her own epistemological model:

> That the truth-value of an observation statement is relative to coherence conditions is a matter of epistemology, but the *concept* of truth that is presupposed is a matter of ontology. . . . Truth is a relation between the state of the world that produces an empirical stimuli and the observation statements expressed in current descriptive language." (Quoted in Potter 1984, 16)

So the ultimate meaning of truth is the traditional one of a relation of correspondence between the knowledge-claim and reality.[9]

With this conception of truth we have a schism drawn between reality and the knower, wherein knowledge claims are expressed by a knower and are about reality. Although our given criterion of truth, coherence, can involve the particularities of the knower or the scientist, the ultimate arbiter of truth-value for any statement is reality. What makes a statement true is not its coherence or its experimental verification but its correspondence to reality. The resulting conception of reality is as an autonomous entity, certainly changing spatiotemporally but not *necessarily* changing between coherent frameworks or fields of belief. This is, of course, our traditional

Western conception of truth and reality, although the epistemic criterion of actual theory-choice has been radically changed. And we should not underestimate the significance of no longer being able to *claim* correspondence for any scientific theory or claim. However, it is unclear to me at least how, with such a conception of truth, values can be conceived as anything but distortions in the epistemic process. Though they cannot be treated differently or even isolated and identified within our actual practice of scientific theory-choice, from a god's-eye-view it would seem that values cannot be epistemically reliable or even relevant within the process of truth-acquisition.

While the Holistic Model scores low on our criterion of integrating values, however, it must, perhaps as a result, score high on our criterion of persuasiveness. For truth in the sense of correspondence is a powerful conception of truth: if one can argue that one has access to such a truth, one's theory will indeed be persuasive. The problem of relativism accordingly recedes. The perplexity, of course, is how a coherentist model of theory-choice can presuppose a correspondence theory of truth without claiming correspondence as a criterion of truth. Conceptually it is certainly possible, but what the significance, and subsequently the advantage, of presupposing correspondence is for the actual practice of theory-choice is far from clear.

The Constructivist Model of theory-choice operates with a very different conception of truth. Whereas correspondence theories posit a truth which is abstract and universal, the Constructivist's truth is historical and contingent. The schism between the knower and the known has been eliminated in such a way that truth can never be free of subjective elements, to use the Holistic terminology. Truth is the product of a process which involves observations, practices, and theories, rather than a correspondence relation between a proposition and a transcendent reality. The process of knowing constitutes what is true, which makes truth something constructed rather than discovered intact. Thus truth is essentially historical and essentially contingent, and it is not simply that our *claims* to truth must be historically contextualized. There is no truth (even an inaccessible truth) which is universal, abstract or independent of the scientific process. Foucault (1972, 49) says, "discourses . . . [are] . . . practices that systematically form the objects of which they speak." He says further that:

> Each society has its regime of truth, its 'general politics' of truth: that is, the types of discourse which it accepts and makes function as true; the mechanisms and instances which enable one to distinguish true and false statements, the means by which each is sanctioned; the techniques and procedures accorded value in the acquisition of truth; the status of those who are charged with saying what counts as true. (1980, 131)
>
> 'Truth' is to be understood as a system of ordered procedures for the pro-

> duction, regulation, distribution, circulation, and operation of statements. (1980, 133)

Similarly for Gadamer truth

> . . . amounts to what can be argumentatively validated by the community of interpreters who open themselves to what tradition 'says to us.' This does not mean that there is some transcendental or ahistorical perspective from which we can evaluate competing claims to truth. We judge and evaluate such claims by the standards and practices that have been hammered out in the course of 'history.'[10]

And for Gadamer just as for Foucault these standards and practices themselves are historical and undergoing constant change and development. So although Gadamer unlike Foucault posits a judging subject as a focal point of the knowledge process, he yet shares with Foucault a notion of truth itself as emerging from a dialogical process rather than being discovered or appropriated as the end-result of a scientific inquiry. There is no conception of a correspondence between statements and autonomous reality laying hidden behind a claim whose justification we can only give as "warrantedly assertable." Truth is about reality, but reality itself is in historical development, an "emergent property" of discourses and practices.

Neither the Constructivist nor the Holistic conception of truth prima facie entails a radical or absolute epistemological relativism. The Holistic conception of truth is correspondence to an independent reality, and so obviously avoids relativism (at least in its *conception* if not its *criterion* of truth). And the Constructivist conception of truth could be construed as positing a unique though contingent truth since truth here is by virtue of real material existents and not the whims of an individual. What is sacrificed by the Constructivist conception is not objectivity but merely universalizability.

Though neither conception prima facie entails radical relativism or rejects (some form of) objectivity, the conceptions differ in two epistemologically important respects. On the Holistic view values must inevitably enter into scientific theory-choice as obstacles or irrelevant additions to truth. If truth is a relation of correspondence to an autonomous reality, subjective, normative beliefs cannot be reflective of that reality (unless one holds a theory of ethical realism such as Brentano or Dewey held). On the Constructivist view, however, values are not obstacles, but are actually constitutive of truth, since truth is an "emergent property" of discourses and practices that involve normative elements.

Secondly, the two conceptions of truth differ in the degree to which they are committed to some form of relativism. On the Holistic view, the *criterion* of theory-choice is relative (to the existing web of belief), although truth itself is not relative. For the Constructivist conception truth itself must be relative to the dialogic process.

In the preceding discussion what is meant by relativism and the kinds and degrees of relativism possible has been left intentionally unclear. I will return to the issue of relativism at the conclusion of this essay. For now, we need to summarize the contribution our explication of these two conceptions of truth can have for the problem of feminist social science and ascertain where that problem now stands.

FEMALE SOCIAL SCIENCE AND THE TRUTH DEBATE

We have seen that in answer to the problem of justifying and explaining feminist social science the relevant literature has offered three models of theory-choice: the positivist model, the Holistic model and the Constructivist model. We have also seen that the distinguishing feature of primary importance between the two latter models is their respective conceptions of truth. Now the question is, where do we go from here?

Unfortunately, none of these models provide an unequivocal and compelling explanation of the sort required. I have rejected the positivist model as the least acceptable because of its commitment to value-neutral theory-choice. So if we are left with a choice between the latter two models, neither of which is conclusively compelling, how are we (prior to accepting a model of theory-choice) to go about choosing between them? In other words, what model of theory choice can we use to choose models of theory choice?

One possible place to begin addressing this perplexing problem would be to trace the implications of the various conceptions of truth for current problems within the research program of feminist social science. Let us look for example at the problem of women's experience. Many feminist theorists have observed that a unique feminist political program, set of interests, perspective and/or epistemological standpoint require grounding in (and thus are justified via) the common experiences women share across class and culture of childbearing and rearing, housework, exclusion from the public domain, etc. It does seem as if to claim the possibility of a politics of women we must be able to articulate and identify the distinguishing characteristics that make up the category 'woman'. And yet the dilemma is that invariably such articulations of women's experience become drawn into an essentialist reading of femininity. And so the debate goes.

To see how our explications of truth can shed light on this controversy, let us look at how the notion of women's experience is problematized depending on which conception of truth we adopt. On the Holistic model of truth as correspondence to an independent reality, women's experience becomes some thing independent (of the process of acquiring it) that we must discover. Catherine MacKinnon's work on the "private sphere" that women share seems to exemplify this kind of approach to the problem of women's experience. She identifies "women's distinctive experience as women" as "within that sphere that has been lived as the personal" (1982, 21; 1983, 635–658). She then goes on to describe and analyze this sphere in a straightforwardly empirical manner to reveal its essential characteristic as a "pervasive powerlessness to men, expressed and reconstituted daily *as* sexuality." Though MacKinnon denies that she is using the scientific method to explain women's experience, her analysis of the material features of women's lives to draw out the truth of that experience belies her claim. MacKinnon's search for the truth of women's experience assumes an autonomous reality in which that experience takes place.

On the Constructivist model of truth as an emergent property of discourses and practices, women's experience becomes something we create, or learn how to create or use, rather than something we simply discover. Teresa de Lauretis's work on the problem of women's experience appears to exemplify this approach. For Lauretis, experience is defined as a "complex of habits resulting from the semiotic interaction of 'outer world' and 'inner world', the continuous engagement of a self or subject in social reality" (1984, 182). The question of women's experience becomes, then, not what are the components of women's lives and relations to others distinctive to them as women, but, what is the process in which women are constructed as subjects? How is it that the material, economic, and interpersonal relations are perceived and comprehended as "subjective (referring to, even originating in) oneself?" (Lauretis 1984, 159). In one sense, Lauretis's questions can be seen as an advance step beyond MacKinnon's, to further theorize and re-problematize the experience that MacKinnon identifies as woman's. In another sense, Lauretis's question can be seen as emanating from a different "research program" (i.e., from a conception of truth as constituted out of a dialogic process). Lauretis does not assume that women's experience *as* constitutive of women exists intact before her pursuit of it, but rather that in the act of conceptualizing and problematizing women as subjects emerges the entity she calls "women's experience."

The difference between MacKinnon and Lauretis is a difference in how they construe the ontology of experience. But *as* an ontological difference, it entails a difference in their epistemological conception of truth, for truth is always defined as a relation with reality and thus always involves one's conception of reality. Thus, MacKinnon's view that experience can be discovered and Lauretis's view that it is constructed through the process of inquiry *follows* from their contrasting conceptions of reality which entails contrasting conceptions of truth, specifically, the Holistic and Constructivist conceptions respectively. Understanding this fundamental level of their difference is crucial to engaging in an effective critique of their approach to the problem of women's experience.

Although the recognition that the difference between MacKinnon's and Lauretis's approach to the problem of women's experience represents a difference in their conceptions of truth will help to clarify the real issues at stake in this debate, this alone cannot, however, decisively settle the debate. It cannot, that is, provide a neutral ground, or ultimate level of arbitration, from which we can observe and ajudicate these confiicting theories of truth.

THE PROBLEM OF RELATIVISM: A FLOATING SIGNIFIER?

Although this essay has served to demonstrate some of the epistemological implications involved in the problem of feminist social science, it could be argued that we are no further advanced toward a solution until the issue of relativism has been clarified. However, it should now be obvious that our original statement of the problem of relativism needs rethinking. Specifically, the concern that theory-choice in feminist social science will lose its emancipatory potential if it entails a radical relativism looks very different depending on which conception of truth we adopt. Does Keller's concern with relativism and ideology presuppose a correspondence theory of truth? Do all worries about relativism presuppose a correspondence theory of truth?

One reason we might think so involves the underlying assumption embedded in the relativist criticism. That is, the charge of relativism serves as a powerful criticism of any theory because it implies that, if our truth is only relative, we have not yet got the real truth. One has a mental conception of a transcendent reality to which our relative truths are only poor or partial approximations. If our beliefs are true only relative to a discourse or a web of belief, then we have reason to suspect that our 'true' beliefs are not very true. This entire line of reasoning, of course, presupposes that the meaning of true (or at least the one that is really important) is correspondence, usually between a proposition and reality.

On the other hand, what happens to the problem of relativism if we drop out from the above mental picture the transcendent, independent reality lying beyond our discourse or web? It seems to me this is what Foucault and Gadamer at least want to do. Our beliefs are still relative to a discourse, but they cannot be characterized as therefore "less true" because they are not being compared to a transcendent reality. This is the move the Constructivist makes in saying that, it is not just the criterion of truth that is relative, but truth itself that is relative. Assuming for the moment that such a view is coherent and tenable, it would appear to make relativism less formidable by undercutting the usual ground its critics take.

However, this move may not help feminists at all. For after all, the emancipatory potential of feminist theory and social science remains an important motivation for research and criterion for methodologies. If we choose a model of theory-choice which undercuts relativism at the same time that it deflates the persuasiveness of our conclusions this emancipatory potential will be lost. We might then be led to ask the converse of our previous question, that is, not does the problem presuppose a correspondence theory of truth, but rather, can the problem be solved without a correspondence theory of truth?'[11]

However, in considering how to make this agenda both efficacious and coherent we should have learned one important lesson from our survey of possible theory-choice models and the underlying epistemological debates they conceal. This is that Evelyn Fox Keller's articulation of the problem is misguided to the extent that it warns against replacing science with ideology. If the defeat of positivism is to be consolidated, we must cease positing such easy distinctions between scientific and political theories and practices. We need not worry about replacing science with ideology because in an important sense science already is ideology, to the extent that political values and commitments already play a significant role in scientific theory-choice. The considered devaluation of science by the incursion of politics presupposes a positivist philosophy of science. The point here is that in articulating possible solutions to the problem of relativism we must be careful to avoid assuming once again that politics and science should not be seen together.

Conclusion

I believe that the problem of feminist social science remains and has not yet been solved by either of the two models of theory choice presented here in their current stage of development. On the other hand, I doubt seriously that we can ever return to positivism. Perhaps the most important and widespread movement throughout Western intellectualism is the move to what Richard Bernstein has characterized as a

> . . . more historically situated, nonalgorithmic, flexible understanding of human rationality, one which highlights the tacit dimension of human judgement and imagination and is sensitive to the unsuspected contingencies and genuine novelties encountered in particular situations. (1983, xi)

The challenge for feminist social science is still to develop a model of theory-choice consistent with this new conception of the process of gaining knowledge without sacrificing its ability to explain and justify its existence as a reliable, though not foolproof, process.

In this paper I have tried to explicate the implicit epistemological debates that are embedded within the theoretical debates in feminist social science. It is my view that talk about feminism and social science needs to become (at least occasionally) talk about feminism and epistemology, and *this* must fundamentally become talk about truth. What do we mean by truth? What can we possibly mean? And further: If we give up positivism and/or a correspondence theory of truth to what theory of meaning can we turn? Until the discussion on these questions begins, the discussions about feminism and social science and feminism and science and even feminist theory generally will (continue to) be stalemated by question-begging and unclarity.

NOTES

I would like to thank Elizabeth Potter, Harriet Whitehead, Richard Schmitt, Vrinda Dalmiya, Denise Riley, Ernest Sosa, James Van Cleve, Susan Haack and the *Hypatia* referees for their kind help on the arguments in this paper. I would also like to thank the participants of the 1984–85 Pembroke Center Seminar on the Cultural Construction of Gender at Brown University for their helpful responses to these ideas and for their encouragement.

1. I do not mean to imply here that politics and science are mutually exclusive, or to devalue political argumentation.

2. I have also heard it argued that those with the least investment in the status quo can somehow see the truth more clearly, but I have never been convinced by this. To the extent that this argument is saying that feminists have less interest in the outcome of their research than male supremacists, surely it is implausible, for while male supremacists may have a lot to *lose*, feminists (and women) surely have a lot to *gain* from the emancipatory implications of their research.

3. In this and in what follows I am posing a distinction between facts and values, but two points need to be kept in mind when these concepts are used. (1) By facts I am not referring to brute data or pure empirical atoms unsullied by theoretical assumptions or predispositions. In my view, all facts are, to widely varying degrees, theory-laden. That is, our observations are mediated by linguistic categories and theoretical commitments. However, there is still a relevant distinction to be made between facts comprising such theory-laden observations and *values*, by which is meant moral and/or political commitments and predispositions. All normative claims are guided by values, in my view, and normative claims can be distinguished from theory-laden descriptive claims or facts. Therefore, when I refer to the question of whether values are involved in empirical claims, I am *already assuming* that theories are so involved. (2) By values I am referring to specific moral and political beliefs of the type "women should have equal opportunities" or "the best society is the most egalitarian society." Hilary Putnam, among others has argued that values are inherent in science by calling criteria like coherence and simplicity *values*. While I completely agree with Putnam's point that such criteria operate as values in science, his argument leaves aside the possibility that the more usual moral and political values are insignificant in scientific theory-choice. Thus, it seems to me that there still exists a type of fact/value distinction which is a legitimate concept and an issue deserving of more discussion. See Hilary Putnam (1971, 127–149).

4. Elizabeth Potter (1984, 10) makes this argument also.

5. My use of the concept "radical relativism" corresponds to Ian Hacking's (1984, 48–66) definition of subjectivism: the view that individuals can exert control over whether a proposition is *true* or not. (Whether this control can be exerted consciously or unconsciously is not relevant here). This concept is opposed to a milder form of relativism in which it is argued that individuals can exert control over whether a proposition is *considered* true, or over whether a proposition is even included among that category of propositions we take as having a truth value at all. For example, the proposition "Zeus is angry" would presumably be a proposition without a truth value, a proposition for which we could say neither that it is true or false.

6. On the incoherence of radical relativism, see Donald Davidson (1984).

7. It is unclear to me whether in her latest work Harding (1986) is moving away from a

Holistic model toward a Constructivist model (which is the second model I will discuss), or whether she is trying somehow to sanction both. In either case, Harding is moving away from a pure Holistic model.

8. For a lucid discussion of this point, see Charles Taylor (1979, esp. 46–50).

9. The issue of whether Quine himself holds or can hold a correspondence theory of truth (given his other commitments) is immensely complex, and turns principally on two issues: whether his adoption of Tarski's semantic theory of truth is compatible with a correspondence theory and whether his doctrine of ontological relativity rules out correspondence. And these two issues turn on which definition or account of the correspondence theory we choose to adopt. There has been philosophical controversy on all these points. For example, both Karl Popper (1962, 1972) and Donald Davidson (1984) have argued that a semantic theory of truth can be combined with a correspondence theory, Popper even claiming that the former provides a more precise explication for the latter. Hartry Field (1980, 83–110) has argued against Popper on this, and Eliot Sober (1982, 369–385) has argued further that the two theories tend away from rather than toward each other. Others have argued that Tarski's theory is epistemologically neutral, e.g. Max Black (1954) and Tarski (1949) himself. Field (1974) has also argued that the doctrine of ontological relativity conflicts with a correspondence theory of truth.

In my view, if correspondence is construed in the loosest possible sense, such that a proposition is made true by virtue of its relation to some extra-linguistic reality rather than in virtue of its relation to other beliefs or to pragmatic success, for example, then it is possible to ascribe a correspondence theory of truth to Quine. As Susan Haack (1978, 113–114) has argued, the satisfaction relation in Tarski's schema of truth can serve as a correspondence relation between a linguistic and an extra-linguistic entity, thus effectively combining the semantic and correspondence theories. However, given his thesis of ontological relativity and the inscrutability of reference, it is misleading to attribute to Quine the view that a particular sentence, if true, corresponds to reality, because any particular sentence will express an ontological commitment that the stimulus on our sensory receptors will vastly underdetermine. Still, in an important sense Quine believes that certain sentences and not others are acceptable because of their connection to sensory stimuli, and not merely or primarily because of their coherence relations. So the set of sentences "A rabbit passed by," "An undetached rabbit part passed by," and "Retinal firing 37 occurred" may each enjoy a relation of correspondence to an extra-linguistic event that other sentences do not, though each relate to the same particular event. It may be that Quine does not wish to confer on this relation the honorific title "true" (see 1981, 39), and yet he does not want to leave our belief in science as merely assertable (see 1983, 24–25). Despite his pragmatic and semantic understandings of how we use the term 'truth' (or perhaps in the midst of these), Quine holds, not unlike Peirce, an empiricist and realist commitment to the relationship, though not the isomorphism, between science and reality, a commitment shared traditionally by the correspondence theory of truth.

For the purposes of the distinction between Holism and Constructivism that I wish to draw in this paper, this loose conception of correspondence is sufficient. The distinction turns on whether the world through perceptual experiences plays a privileged part in determining truth, or whether human practice, theoretical commitment, and theoretically interpreted observations contribute equally to the construction of truth. Note that in the first case truth is discovered and in the second it is constructed. In light of the free reign Quine accords science, he may be attributed with a view that mediates these two extremes, but in no case given his physicalist commitments can Quine be labeled a pure Constructivist.

10. This is Richard Bernstein's (1983, 154) useful and I believe accurate encapsulation of Gadamer's conception of truth. See Gadamer (1984, especially second part).

11. My gratitude goes to Vrinda Dalmiya for helping me to see this aspect of the problem.

References

Bernstein, Richard. 1983. *Beyond objectivism and relativism: Science, hermeneutics, and praxis*. Philadelphia: University of Pennsylvania Press.

Black, Max. 1954. The semantic conception of truth. In *Philosophical analysis*, ed. M. MacDonald. New York: Basil Blackwell.

Davidson, Donald. 1984. *Inquiries into truth and interpretation*. Oxford: Clarendon Press.

Duhem, Pierre. 1962. *The aim and structure of physical theory*. Trans. Philip P. Wiener. New York: Atheneum Press.

Dye, Nancy Schrom. 1979. Clio's American daughters: Male history, female reality. In *The prism of sex: Essays in the sociology of knowledge*, ed. Julia Sherman and Evelyn Torton Beck. Madison: University of Wisconsin Press.

Eichler, Margaret. 1980. *The double standard: A feminist critique of feminist social science*. New York: St. Martin's Press.

Field, Hartry. 1974. Quine and the correspondence theory. *Philosophical Review* April: 200–228.

———. 1980. Tarski's theory of truth. In *Reference, truth and reality*, ed. Mark Platts. Boston: Routledge and Kegan Paul.

Foucault, Michel. 1972. *The archaeology of knowledge*. Trans. A. M. Sheridan Smith. New York: Pantheon.

———. 1980. *Power/knowledge: Selected interviews and other writings*. Ed. Colin Gordon. Trans. Colin Gordon et al. New York: Pantheon.

Gadamer, Hans-Georg. 1976. *Philosophical hermeneutics*. Trans. David E. Linge. Berkeley: University of California Press.

———. 1984. *Truth and method*. New York: Crossroad Publishing Company.

Haack, Susan. 1978. *Philosophy of logics*. Cambridge: Cambridge University Press.

Hacking, Ian. 1984. Language, truth and reason. In *Rationality and relativism*, ed. Martin Hollis and Steven Lukes. Cambridge, MA: MIT Press.

Harding, Sandra, ed. 1976. *Can theories be refuted?* Boston: Reidel.

———. 1977. Does objectivity in social science require value-neutrality? *Soundings* LX (4): 351–366.

———. 1986. *The science question in feminism*. Ithaca, NY: Cornell University Press.

Harding, Sandra and Merrill B. Hintikka, eds. 1983. *Discovering reality*. Dordrecht, Holland: D. Reidel.

Hartsock, Nancy. 1983. *Money, sex and power*. New York: Longman Publishing Company.

Hesse, Mary. 1976. Duhem, Quine, and a new empiricism. In *Can theories be refuted?* See Harding 1976.

———. 1980. *Revolutions and reconstructions in the philosophy of science*. Bloomington, IN: Indiana University Press.

Hubbard, Ruth. 1983. Have only men evolved? In *Discovering reality*. See Harding and Hintikka 1983.

Keller, Evelyn Fox. 1982. Feminism and science. In *Feminist Theory*, eds. Nannerl O. Keohane et al. Chicago: University of Chicago Press.

———. 1985. *Reflections on gender and science*. New Haven: Yale University Press.
Kuhn, Thomas. 1970. *The structure of scientific revolutions*. Second edition. Chicago: University of Chicago Press.
Lauretis, Teresa de. 1984. *Alice doesn't*. Bloomington, IN: Indiana University Press.
MacKinnon, Catherine. 1982. Feminism, marxism, method, and the state: An agenda for theory. See Keller 1982.
———. 1983. Feminism, marxism, method and the state: Toward feminist jurisprudence. *Signs* 8 (4): 635–658.
Peirce, C. S. 1940. *The philosophy of Peirce: Selected writings*. Ed. Justus Buchler. New York: Harcourt, Brace and Company.
Popper, Karl. 1962. On the sources of knowledge and inference. In *Conjectures and refutations*. New York: Basic Books.
———. 1972. *Objective knowledge*. Oxford: Oxford University Press.
Potter, Elizabeth. 1984. Feminism and the crisis of objectivity. Paper presented at the American Philosophical Association conference in New York, December.
Putnam, Hilary. 1971. *Reason, truth and history*. Cambridge: Cambridge University Press.
Quine, W. V. O. 1963. Two dogmas of empiricism. In *From a logical point of view*. New York: Harper and Row.
———. 1981. *Theories and things*. Cambridge, MA: Harvard University Press.
———. 1983. *Word and object*. Cambridge, MA: MIT Press.
Reed, Evelyn. 1978. *Sexism and science*. New York: Pathfinder Press.
Smith, Dorothy. 1977. Some implications of a sociology for women. In *Woman in a man-made world*, ed. Nona Glazer and Helen Youngelson Waehrer. Second edition. Chicago: Rand McNally College Publishing.
Sober, Eliot. 1982. Realism and independence. *Nous* XVI (3): September: 369–385.
Stanley, Liz and Sue Wise. 1983. *Breaking out: Feminist consciousness and feminist research*. Boston: Routledge and Kegan Paul.
Tarski, Alfred. 1949. The semantic conception of truth. In *Readings in philosophical analysis*, ed. H. Feigl and W. Sellars. New York: Appleton-Century-Crofts.
Taylor, Charles. 1979. Interpretation and the sciences of man. In *Intepretive social science*, ed. Paul Rabinow and William M. Sullivan. Berkeley: University of California Press.
Westkott, Marcia. 1979. Feminist criticism of the social sciences. *Harvard Educational Review* 49(4): November: 422–430.

John Dewey and Evelyn Fox Keller: A Shared Epistemological Tradition

LISA HELDKE

In this paper, I undertake an exploration of the similarities I find between the epistemological projects of John Dewey and Evelyn Fox Keller. These similarities, I suggest, warrant considering Dewey and Keller to share membership in an epistemological tradition, a tradition I label the "Coresponsible Option." In my examination, I focus on Dewey's and Keller's ontological assertion that we live in a world that is an inextricable mixture of certainty and chance, and on their resultant conception of inquiry as a communal relationship.

John Dewey and Evelyn Fox Keller undertake epistemological projects that bear a striking number of similarities to each other. The two share beliefs about the nature and aims of the activity of inquiry, the nature of the inquiry relationship, and the composition of the community in which inquiry takes place. The similarities between their projects make it possible, and, I'd claim, fruitful, to regard them as members of the same epistemological tradition. This paper is an exploration of a few of these deep and important similiarities.

To begin, I'll give a thumbnail sketch of the tradition to which I claim Keller and Dewey belong. This sketch will provide a general context into which my discussion of particular aspects of their projects may be placed. I call this tradition the "Coresponsible Option," a name that hints at several aspects of its composition.[1] First of all, the term "coresponsible" evokes the communal aspect of inquiry. Inquiry, as we shall see in more detail later, is an activity that takes place between two "things" that have *responsibilities* to each other, obligations to treat each other with respect and care.[2]

I use the word "option" in the name of this tradition to emphasize its provisional nature. This "provisionality" has two aspects. First, coresponsible theorists seek to undermine the absolutist claim that theories must ultimately be grounded in some "real, independent, antecedent world," or in some set of "purely rational principles."

Coresponsible positions reject this absolutism, but they also reject the

Hypatia vol. 2, no. 3 (Fall 1987). © by Lisa Heldke.

relativist alternative—that there are no absolute grounds for knowledge, therefore no grounds of any sort for knowledge. Rather than choosing either of these alternatives, coresponsible theorists claim that we can indeed construct provisional grounds, provisional foundations, for our knowledge. Provisional grounds are claims, beliefs, theories we take to be useful and reliable, but whose obsolescence we can always imagine. Provisional grounds are useful "springboards" for future investigation, but they cannot, in principle, ever be regarded as "ultimate." Modification in light of new evidence is a part of their very *nature*; their usefulness derives in part from their fluidity and changeability.

In the same spirit, I'd suggest that the entire tradition is itself an option; it is one useful epistemological project, not the only useful one. It is a tradition whose acceptance is invited, not demanded. I offer the Coresponsible Option in the spirit of a cook who passes out copies of favorite, well tested recipes. It has proven reliable for me, and I'm willing to claim that it will work equally well in other contexts. But, as with recipes, I'd argue that no epistemological program works in all contexts for all users. Maybe you're cooking under high altitude conditions. Or maybe you just don't like tripe.

With this highly abbreviated sketch of the Coresponsible Option, I leave my discussion of the tradition-in-general, and turn to two of its members, Dewey and Keller. My exploration of them will focus on their conceptions of the nature and aims of inquiry, and on their formulations of the inquiry relationship.

Two general aims motivate my examination here. First, I wish to legitimate my claim that Dewey and Keller are in fact members of the same epistemological tradition—that important structural connections exist between their two projects. Second, I wish to commend these projects as useful ways to free ourselves from the dualistic and hierarchical binds into which many traditional epistemologies place us.

I begin with a brief discussion of ontology. Dewey's and Keller's assumptions about the nature of the world—a world that is both certain and uncertain—are instrumental in the formation of their views of the nature of inquiry, so it is important that we first consider them.

One striking feature of their ontological views is the attention they both pay to the element of chance in the world. Both emphasize the importance of recognizing that the world we inhabit is an inextricable *compound* of stability and change, chance and certainty; that *both* are obvious and significant aspects of our actual experience in the world. Their belief that chance, difference, anomaly, are important parts of our experience, just as are stability, certainty, normalcy, stands in sharp contrast to the bulk of our philosophical tradition, where the certain and stable are regarded as ontologically superior and espistemologically more valuable. Some philosophers in this tradition have argued that only the unchanging is knowable. That which changes, these philosophers argue, cannot be known; and therefore it cannot be as important, either. Indeed, some philosophers have gone so far as to claim that only the stable and certain are "real"; things that change have merely an illusory or "reflectional" existence.

Dewey, in *The Quest For Certainty* ([1929] 1980), suggests that the history of western philosophy might be read as the history of an attempt to establish absolute certainty in nature, where no such certainty is given to our experience. It is a record of attempts to abolish, to deny the existence of, the changing and the uncertain. Traditionally, he notes, philosophers have posited a hierarchical bifurcation between chance and certainty, change and permanence. At times they have even denied the reality of chance and change.

Dewey suggests that our fear of the unknown, our desire for stability and security in the face of chaos and unexpectedness, might be the forces behind these attempts. These fears and desires have caused us wishfully to deny the reality of uncertainty, and to claim that the "true" nature of the world is wholely knowable and predictable.

But, he counters, such "wishful thinking" cannot cover up the fact that our experience in the world *is* always experience of certainty and change commingled. When we look to experience, we see that we live in a world that's an inextricable compound of stability and change. We are misguided if, " . . . when we pull out a plum, we treat it as evidence of the real order of cause and effect in the world" ([1929] 1980, 45).

We cannot, in fact, conceive the nature of one part of this compound without thinking of it in relation to the other. Even the word "compound" fails fully to explicate the interrelation between chance and certainty, for it suggests that the component parts are actually independent elements that may exist and be defined in isolation from one another. But in truth, we conceive of, define and experience the compound, and only derivatively experience its constituent "parts." Dewey asserts that the two " . . . are mixed not mechanically but vitally like the wheat and tare of the parable . . . [C]hange gives meaning to permanence, and recurrence makes novelty possible" ([1929] 1980, 47).

Traditional western philosophy, as I have said, has often portrayed knowledge as the grasping of something that does not change. On this view, it is imperative, if knowledge be possible, for something unchanging to exist. This conception of knowledge, and of what can be known (arising, as Dewey suggests, out of our desperate fear of the unknown, and our desire for security) results in our either ignoring chance and change, or assigning them an inferior, derivative ontological and epistemological status. But, Dewey suggests, "the world must actually be such as to generate ignorance and inquiry; doubt and hypothesis; trial and temporal conclusions . . . " (1958, 69). We do inhabit a world of genuine chance and stability; and our inquiry into it must take account of that situation and provide us with means for coping—flourishing—in the face of it.

Dewey's ontological assertion of the "genuineness" of the chance-and-certainty compound helps to shape his theory of inquiry. It is a theory that emphasizes knowing, the activity, over knowledge, the state. Knowing, for him, is the task of building stability in a not-altogether-stable world. It is a process of living with, and utilizing, instability. Its aim is not the construction of a body of fixed knowledge for contemplation but the development of a flexible body of ways to live with change.

Knowledge, the static body of "data" that is the "property" of some knower, is useful and interesting insofar as it allows the knower to "go on"; to ask the next question, to solve the next problem, to relieve some anxiety, to throw over the whole project and assume a different tack. Inquiry proceeds in an atmosphere in which objects of knowledge always and in principle retain their provisional character (1958, 130). Knowledge, in a sense, works in the "service" of knowing. It is the temporary firm spot from which inquiry can set off, and to which it can return, for confirmation or refutation. And of course it is always subject to change in the light of new evidence.

It should be noted that Dewey does not deny the existence of knowledge altogether; his criticism is of the notion of permanent knowledge as an end in itself. He doesn't question the existence of, say, contemplative thought, but the all-importance of it.

Related to his emphasis on knowing as an activity is Dewey's assertion that knowing is ultimately practical. He disavows the traditional hierarchical distinction between theory and practice, a distinction in which theory, "purely mental" activity is associated with knowledge (the armchair-and-smoking-jacket "thinker"). Theory, on this picture, is thus regarded as epistemologically superior to practice, which is merely practical, and aims only at coping with the uncertainty inherent in everyday life. Practice ranges over the "ordinary" world: "Instead of being extended to cover all forms of action by means of which all the values of life are extended and rendered more secure . . . , the meaning of 'practical' is limited to matters of ease, comfort, riches, bodily security and police order, possibly health, etc. . . . In consequence, these subjects . . . are no concern of 'higher' interests which feel that no matter what happens to inferior goods in the vicissitudes of natural existence, the highest values are immutable characters of the ultimately real" (1980, 32). Dewey erases both dichotomy and hierarchy between theory and practice, by expanding the scope of practice to include theory, and then ascribing to it an epistemological status befitting its importance.[3] The difference between theory and practice is one of degree, not kind; a distinction between " . . . two modes of practice. One is the pushing, slam-bang, act-first-and-think-afterwards mode, to which events may yield as they give way to any strong force. The other mode is wary, observant, sensitive to slightest hints and intimations" (1958, 314).

I turn now to Keller's ontology, to consider how it grows from and informs her understanding of the inquiry activity. Keller, like Dewey, rejects the inherited dichotomy of chance and certainty. In her work, she stresses the relevance to inquiry of the anomalous event, and emphasizes the need to understand the anomaly and the expected event in relation to each other.

Keller uses the example of geneticist Barbara McClintock, whose work exemplifies Keller's conception of an interweaving interdependency between chance and certainty. Her work also illustrates the need for responsible inquirers to be sensitive to this interdependency. The world we experience is both changing and stable; it will not do to construct a system that denies the reality of one, or that treats the two as separate facets.

This conception of the nature of the world we experience informs and is informed by Keller's notion of the nature of inquiry. Again, let's look to the example of Barbara McClintock. McClintock, Keller suggests, sees her own work as a means by which to make difference—the anomalous, the unexpected event with which nature confronts us—understood. McClintock attempts to weave difference into the fabric of theory in a way that allows it to retain its uniqueness, but that locates that uniqueness within a context, a continuity in nature. The goal of inquiry, contrary to traditional epistemological thinking, is not to make difference disappear within sameness, or to attempt to subsume the chance event under the law of the expected outcome, but to render difference understandable, articulable on its own terms. Chance events, anomalous events, have meanings of their own, not reducible to the meaning of the certain. Exceptions don't prove or refute laws of nature; they are significant in their own right, and in relation to laws of nature. Difference *as* difference may teach us some things.

This respect for difference works alongside certainty and sameness, not in opposition to it. Dewey, we saw, escapes the dichotomization of chance and certainty by understanding their essential interdependence. Keller similarly asserts that difference is connected to sameness, that the two are interdependent. Inquiry itself begins in a recognition of difference, but not in a division, a bifurcation, a hierarchical ordering of certain over uncertain, uniform over exceptional. For it is only within the context of the orderly world that the richness of the anomaly can be realized—and it is only in the company of the anomaly that the full significance of order can be recognized. McClintock, in theorizing, seeks to weave together the strands of her experience in a manner that allows each to retain its integrity while giving it a relationship to order.

In an attempt to incorporate this notion of the relationship of anomaly to the rest of the world, Keller enlists a concept she calls "order." She suggests conceiving of nature as "ordered" rather than as "lawful," because, as she envisions it, order is a broader, more flexible category than law, and it is one that allows us to see nature as neither inherently "bound by law" (i.e., absolutely certain) nor "chaotic and unruly" (i.e., entirely chance). "Order" encompasses both modes of organization that arise "from within" phenomena themselves—that grab us by the neck and say "understand us this way"—and those that are imposed from without—that we place upon the phenomena (1985 133–134). With this description, Keller, like Dewey, impresses upon us the need to see the relationship between chance and certainty not as a dichotomy, or a hierarchy, but as a relationship of mutual interdependence, of "interdefinition." In "order," Keller evinces the spirit of a world in which much of what we experience *may* be explained by law-typically-conceived, but in which significance may also derive from the exception, the anomaly, the chance event.

Gender plays an important role, both critical and constructive, in Keller's analysis of the relationship of chance and certainty. Dewey's analysis and critique of the dualism is couched in an examination of the history of western philosophy and of a general human desire for stability. But he does not explore how particular human conditions might have been instrumental in shaping the dualism. Keller, in her analysis of gender, does so. And in her discussion of McClintock, she shows how gender may be used to overcome that dualism.

Barbara McClintock is a woman working in a profession long and largely dominated by men. What may we learn from her about how a consideration of gender may help us to construct ontologies that do not adopt the dichotomy of chance and certainty?

In response to such a question, Keller would be quick to point out that, while hardly a "typical scientist," McClintock is not a "typical woman" either. She is an outsider to virtually all the worlds she inhabits. But, setting this disclaimer aside, " . . . however atypical she is as a woman, what she is not is a man. [And] . . . as McClintock herself admits, the matter of gender never drops away" (1985, 174).

McClintock is a woman-in-a-man's-world, in a sense of that expression that plunges far below surface demographic facts about geneticists. Keller argues that McClintock, a "non-man," works within an institution whose basic principles have been greatly influenced by issues of gender—specifically by their having been established by men in cultures where men occupy a dominant position. And, as a non-man, " . . . her commitment to a gender-free science has been binding; because concepts of gender have so deeply influenced the basic categories of science, that commitment has been transformative" (1985, 174). McClintock, given her commitment to being a "scientist," not a "lady scientist," could not but come up with new categories for viewing her world. The old categories simply would not allow her to exist; they would make her out to be an enigma, something to be explained away because of her differentness. But McClintock, in her life as well as in her science, shows that it is possible to make use of difference without demanding its radical separation from or disappearance into, the norm. Her desire to " . . . claim science as a human rather than a male endeavor" gives rise to this view of difference: "It allows for the kinship that she feels [to the extent that she does] with other scientists, without at the same time obligating her to share all their assumptions" (1985, 175).

McClintock is a piercing example of a woman (who does not claim to be a feminist) performing in the patriarchal institution of science, in a way that challenges that institution's patriarchal nature. Her "philosophy of inquiry"—to treat difference as understandable in a context, but also as significant in its own right—is also a statement of her wish for the behavior of the scientific community; namely, to respect the differences of its practitioners, to allow those differences to "speak" in the context of the community, not to stifle or ostracize them. McClintock's own life in the western scientific community is a model of the relation she sees obtaining between the anomalous and the law-governed event in the world of the corn plants she studies.

Moving now to Keller's conception of the nature of inquiry we find that she, like Dewey, emphasizes the importance of *knowing,* the activity. This is hardly surprising, for, if we respect difference, we recognize that the goal of unified, static knowledge is an illegitimate goal. A respect for difference " . . . remains content with multiplicity as an end . . . ," and, I'd add, with the activities of modification, revision, alteration. The activities of posing solutions, plans of action, ways of approaching a situation, are valuable in their own right, in their activity.

Furthermore, Keller's conception of order recognizes that many different ways of organizing events may be used effectively—and not just the many varieties that are produced by using artistic rather than scientific rather than sociological inquiry techniques, but the varieties that are engendered by realizing that many adequate explanations of a given phenomenon may arise in a single branch of inquiry, depending upon where one looks for the source of one's organizing principle.

I turn now to examine Dewey's and Keller's conceptions of the inquiry relationship. My starting point for this investigation will be Dewey's view of the nature of the world in which and of which we are inquirers.

Dewey asserts that the world we observe, inquire into, theorize about, is a world we are *in*, a world that comes to be through our communication with it. The world is not an "external" world for him; it contains humans and their inquiry just as surely as it contains deciduous trees and their leafing cycles. To posit our existence as separate from the world and then wonder how we can ever have knowledge of it is to fabricate a problem where none exists. To regard it as odd that we can "know things outside ourselves" is as silly and, well, downright mistaken, as to find it odd that we can *eat* things external to ourselves. We are not disinterested, detached spectators, but participants in a world that has been and is being created through the activity of all its constituent parts—inquirers and things inquired into.[4]

This view of our place within the world we investigate informs Dewey's conception of the relation between the inquirer and the inquired. Through the activity of inquiry, the natures of both change when, for example, humans come to know trees. The things on both "ends" of this relation are altered through their interaction with each other. In fact, human inquiry *into* the world might more properly be termed interaction *within* it. It is like " . . . the correspondence of two people who 'correspond' in order to modify one's own ideas, intents and acts . . . " (1958, 283). Inquiry in the world is simply one—albeit quite specialized—kind of interaction between organisms.

To be more specific, Dewey treats inquiry as a relationship, a dialectical relationship between inquirer and inquired. Gone is the glass wall that separates inquiring "subject" from inquired-into "object." On traditional models, the subject stands behind this wall, thereby eliminating the possibility of interfering with their object of study, and also ensuring that they will remain untouched by it. Dewey argues that, in fact, inquiry involves precisely the type of mutual influence that the traditional accounts wish to deny.

Knowing, for Dewey, is quite literally communication, both with other inquirers, and with the things inquired into. Neither the concept of creation nor that of discovery adequately captures the Deweyan notion of inquiry—for both these categories assume a subject/object relationship. But for Dewey, one's perspective as inquirer is not that of disinterested, removed, epistemologically privileged bystander, but of dynamic participant, engaged in reconstructing our understanding of, and relationships in, the world through our communication with and in it.

The world in which this inquiry goes on is not an infinitely-malleable piece of silly putty that may be molded however inquirers like; this is not what it means to say that it comes to be through our inquiry into it. It is, rather, a flexible fabric of interconnections, connections capable of being described and developed from myriad perspectives within the world. Humans are part of this fabric; their inquiries are one species of interconnection, one sort of description and development.

This conception of inquiry as communication also lends credence to Dewey's assertion of the importance of inquiry, the activity: Knowing constantly turns up new ways to look at things, new kinds of things at which to look. We are constantly reminded that it's not the task of inquiry to lay down answers, but rather to open up new paths of discourse, to reveal new ways to deal with situations, new kinds of connections in the world. Conclusions are always provisional, and various solutions may coexist without cancelling out the validity of each other.

Keller's model of the inquiry relation is strikingly similar to Dewey's, in which we and the world communicate with and shape each other. She too stresses the dynamic property of inquiry, and rejects the subject/object model of the relationship. The world does not simply present itself to inquirers, or force itself upon us, fully categorized, but neither do we do all the "construction work" ourselves. In different terms, neither creation nor discovery describes our activity. Rather, both ends of the inquiry relation are active participants in the process of inquiry; both ends "tell" each other and are "told" by each other how things work. Keller in fact terms the relationship between inquirer and inquired "dynamic objectivity." The name suggests the aspects of flexibility and reciprocity she regards as essential to the nature of inquiry.

It is necessary, Keller suggests, to recognize that the things into which we inquire are *different* from us, while at the same time we must not sever inquirer from inquired. Just as the presence of anomaly in nature should be seen not simply as an exception to a rule, but as significant because of its difference and productive of a richer picture of the world, so too should we recognize the obvious " . . . difference between self and other as an opportunity for a deeper and more articulated kinship. The struggle to disentangle self from other is itself a source of insight—potentially into the nature of both self and other" (1985, 117).

Of course Keller would not suggest that all inquiry relationships involve the same connections between inquirer and inquired—she does not argue that the relationship between, say, Barbara McClintock and a particular strain of corn is precisely like the relationship between McClintock and one of her colleagues. What she does maintain is that certain very significant aspects of the relationship between McClintock and her corn will be lost, masked, denied by using the hierarchical, dualistic model of inquiring subject/inquired-into object. Specifically, we run the risk of failing to see the rest of the world as comprising individuals with integrity, with things to "say for themselves," and instead regard it as a set of tools-for-our-use.

Dynamic objectivity " . . . aims at a form of knowledge that grants to the world around us its independent integrity but does so in a way that remains cognizant of, indeed relies on, our connectivity with that world" (1985, 117). Dewey describes this relationship as a form of friendship; Keller calls it a form of love: " . . . the scientist employs a form of attention to the natural world that is like one's ideal attention to the human world. . . . The capacity for such attention, like the capacity for love and empathy, require a sense of self secure enough to tolerate both difference and continuity . . . " (1985, 117–118).

Clearly, the omnipotent subject would fail at this sort of inquiry relationship; such a subject demands "submission" from the object of inquiry in order for a relationship even to be established. But our world is not one to be subdued; it is one to learn from and learn with. Inquiry is not a battle, it is communication. It is a relation not unlike love.

To one who argues for the necessity of assuming a detached, disinterested, neutral position from which to perceive the "actual" nature of the world, Keller might respond, "but there isn't one right way to describe nature, and to suggest that there is, to pretend that all you are doing is acting as an amanuensis for nature as it tells you its story, is to misrepresent your activity." Inquiry is a project undertaken in the world to construct and install sets of ordering principles, principles that are colored by one's epistemological values and aims. To feign neutrality and "non-intervention" in inquiry is to mask the fact that we, as inquirers, place laws on nature. It is to attempt to hide the traces of our own biases under a mask of claims to neutrality, rationality and objectivity. It is to fall utterly to understand what inquiry *is*; a process of communication, not of classification.

But, although Keller rejects this "spectator" theory of knowledge, this is not to say that she adopts the alternative idea that all ordering is human ordering, that the world pipes up and tells us nothing. Nature constantly challenges this picture of our role, by handing us events that cannot be fitted into our accepted systems of rules, but that cannot be ignored either.

On Keller's view of the relationship between inquirer and inquired, it is indeed true that we describe "actually existent" features of our world with scientific laws—say, Boyle's law. "But it is crucial to recognize that [this law] is a statement about a particular set of phenomena, prescribed to meet particular interests and described in accordance with certain agreed-upon criteria of both reliability and utility" (1985, 11). Boyle's law is useful and successful, and that success is rooted in the fact that it describes a set of actually-experienced phenomena. But those phenomena were selected by Boyle, and by his scientific tradition, to be the relevant ones. They did not present themselves as the only candidates; rather, over time, they came to be the candidates of choice. Boyle's tradition had taught him to ask certain kinds of questions. His vantage point—his relation to the phenomena he described—was one he inherited from a long, respected and successful scientific tradition. The phenomena he saw and codified were those his tradition brought him to see. So it's not at all a matter of claiming that gases "don't behave that way"—or that of course they behave that way because we make them do so—but rather of realizing that we've decided to, and have been brought up to, pay attention to certain aspects of phenomena and to ignore others. Asking different, but equally legitimate, questions might well produce different results—recall the concept of order. The scientist's world may be described with "realistic" accuracy, but it is the scientist who decides how far that world extends.

Now, to sum up. Dewey and Keller, I've argued, are up to the same things;

they both undertake epistemological projects that are, in general, anti-dualistic, and anti-hierarchical. For both, the world in which inquiry goes on is a world of chance-and-certainty, anomaly-and-law-governed event. This ontological conviction results, for both, in epistemologies that emphasize knowing-as-activity over knowledge-the-state (or contents of state), and that recognize the importance of practical, as opposed to "purely theoretical" activity.

The most significant difference in their views of the nature of inquiry regards the way they come to those views. For Dewey, they arise from an examination of history—primarily the history of western philosophy. He devotes relatively little attention to an examination of how that history was shaped by human conditions—economic, social, sexual, racial—nor does he use these conditions to construct his coresponsible position. Keller, on the other hand, places great emphasis on them. Specifically, she explores the role gender has played in forming our tradition, and shows how a consideration of gender may help us to get out from under that tradition, and formulate a coresponsible position.

Dewey and Keller also restructure the inquiry relationship, rejecting the received, subject/object model, in all its dualistic hierarchy, and replacing it with a "community" model, in which inquirer and inquired alike influence each other. If there is a significant difference between their views here, it lies not in the position each takes, but in the particular aspects of the position each chooses to emphasize. Dewey concentrates on the ways inquirers do affect the course of nature, on the fact that the kinds of interconnections existing in the world do change as a result of our inquiry into them. Keller pays particular attention to the claim that in inquiry, it is important for inquirers to remember both their connection to and their separation from the things into which they inquire. But both agree that inquiry is a connecting activity, connecting inquirer and inquired in myriad ways.

NOTES

1. Thanks to Stephen Kellert for suggesting this name to me.

2. The term "coresponsible" has another significance as well, which I will mention, although I won't discuss it here. It is this: epistemological projects in this tradition subscribe to a *kind* of correspondence theory of truth, although not to the correspondence theory as traditionally conceived. Dewey, for example, derives the notion of truth from the more fundamental notions of usefulness and benefit; he defines it as " . . . processes of change so directed that they achieve an intended consummation" (1958, 161). This definition clearly bears little similarity to the notion that "true" describes some one-to-one correspondence between belief and fact.

All the same, there is a sense in which projects in this tradition utilize a "correspondence" theory of truth. It is the sense in which Dewey speaks of inquiry as a "correspondence" between two friends. On this account, true statements, true theories, are those that arise from successful "correspondence"—successful inquiry.

Furthermore, coresponsibility carries with it the idea of responsiveness. There are no meanings—and there is no truth—independent of interaction. There's no correspondence without response.

3. Here Dewey might be said to "presage" a number of feminists. His recognition of the importance of the practical, of its continuity with theory, is one step toward their theoretical emphasis on bodily activity, especially activity in the home.

4. It might be worth mentioning that the categories of inquirer and inquired are not fixed. One is not locked into one role or the other, even within the context of a given inquiry situation.

REFERENCES

Dewey, John. 1958. *Experience and nature*. 2nd ed. New York: Dover.

———. [1929] 1980. *The quest for certainty*. New York: Perigee.

Keller, Evelyn Fox. 1985. *Reflections on gender and science*. New Haven: Yale University Press.

Part Three

Feminist Critiques of the Practice of Science

Science, Facts, and Feminism

RUTH HUBBARD

Feminists acknowledge that making science is a social process and that scientific laws and the "facts" of science reflect the interests of the university-educated, economically privileged, predominantly white men who have produced them. We also recognize that knowledge about nature is created by an interplay between objectivity and subjectivity, but we often do not credit sufficiently the ways women's traditional activities in home, garden, and sickroom have contributed to understanding nature.

The Facts of Science. The Brazilian educator, Paulo Freire, has pointed out that people who want to understand the role of politics in shaping education must "see the reasons behind facts" (Freire 1985,2). I want to begin by exploring some of the reasons behind a particular kind of facts, the facts of natural science. After all, facts aren't just out there. Every fact has a factor, a maker. The interesting question is: as people move through the world, how do we sort those aspects of it that we permit to become facts from those that we relegate to being fiction—untrue, imagined, imaginary, or figments of the imagination—and from those that, worse yet, we do not even notice and that therefore do not become fact, fiction, or figment? In other words, what criteria and mechanisms of selection do scientists use in the making of facts?

One thing is clear: making facts is a social enterprise. Individuals cannot just go off by themselves and dream up facts. When people do that, and the rest of us do not agree to accept or share the facts they offer us, we consider them schizophrenic, crazy. If we do agree, either because their facts sufficiently resemble ours or because they have the power to force us to accept their facts as real and true—to make us see the emperor's new clothes—then the new facts become part of our shared reality and their making, part of the fact-making enterprise.

Making science is such an enterprise. As scientists, our job is to generate facts that help people understand nature. But in doing this, we must follow rules of membership in the scientific community and go about our task of fact-making in professionally sanctioned ways. We must submit new facts to review by our colleagues and be willing to share them with qualified strangers by writing and speaking about them (unless we work for private companies with proprietary interests, in which case we still must share our facts, but

Hypatia vol. 3, no. 1 (Spring 1988).

only with particular people). If we follow proper procedure, we become accredited fact-makers. In that case our facts come to be accepted on faith and large numbers of people believe them even though they are in no position to say why what we put out are facts rather than fiction. After all, a lot of scientific facts are counterintuitive, such as that the earth moves around the sun or that if you drop a pound of feathers and a pound of rocks, they will fall at the same rate.[1]

What are the social or group characteristics of those of us who are allowed to make scientific facts? Above all, we must have a particular kind of education that includes graduate, and post-graduate training. That means that in addition to whatever subject matter we learn, we have been socialized to think in particular ways and have familiarized ourselves with that narrow slice of human history and culture that deals primarily with the experiences of western European and North American upper class men during the past century or two. It also means that we must not deviate too far from accepted rules of individual and social behavior and must talk and think in ways that let us earn the academic degrees required of a scientist.

Until the last decade or two, mainly upper-middle and upper class youngsters, most of them male and white, have had access to that kind of education. Lately, more white women and people of color (women and men) have been able to get it, but the class origins of scientists have not changed appreciably. The scientific professions still draw their members overwhelmingly from the upper-middle and upper classes.

How about other kinds of people? Have they no role in the making of science? Quite the contrary. In the ivory (that is, white) towers in which science gets made, lots of people are from working class and lower-middle class backgrounds, but they are the technicians, secretaries, and clean-up personnel. Decisions about who gets to be a faculty-level fact-maker are made by professors, deans, and university presidents who call on scientists from other, similar institutions to recommend candidates who they think will conform to the standards prescribed by universities and the scientific professions. At the larger, systemic level, decisions are made by government and private funding agencies which operate by what is called peer review. What that means is that small groups of people with similar personal and academic backgrounds decide whether a particular fact-making proposal has enough merit to be financed. Scientists who work in the same, or related, fields mutually sit on each other's decision making panels and whereas criteria for access are supposedly objective and meritocratic, orthodoxy and conformity count for a lot. Someone whose ideas and/or personality are out of line is less likely to succeed than "one of the boys"—and these days some of us girls are allowed to join the boys, particularly if we play by their rules.

Thus, science is made, by and large, by a self-perpetuating, self-reflexive group: by the chosen for the chosen. The assumption is that if the science is "good," in a professional sense, it will also be good for society. But no one and no group are responsible for looking at whether it is. Public accountability is not built into the system.

What are the alternatives? How could we have a science that is more open and accessible, a science *for* the people? And to what extent could—or should—it also be a science *by* the people? After all, divisions of labor are not necessarily bad. There is no reason and, indeed, no possibility, that in a complicated society like ours, everyone is able to do everything. Inequalities which are bad, come not from the fact that different people do different things, but from the fact that different tasks are valued differently and carry with them different amounts of prestige and power.

For historical reasons, this society values mental labor more highly than manual labor. We often pay more for it and think that it requires more specifically human qualities and therefore is superior. This is a mistake especially in the context of a scientific laboratory, because it means that the laboratory chief—the person "with ideas"—often gets the credit, whereas the laboratory workers—the people who work with their hands (as well as, often, their imaginations)—are the ones who perform the operations and make the observations that generate new hypotheses and that permit hunches, ideas, and hypotheses to become facts.

But it is not only because of the way natural science is done that head and hand, mental and manual work, are often closely linked. Natural science requires a conjunction of head and hand because it is an understanding of nature *for use*. To understand nature is not enough. Natural science and technology are inextricable, because we can judge that our understanding of nature is true only to the extent that it works. Significant facts and laws are relevant only to the extent that they can be applied and used as technology. The science/technology distinction, which was introduced one to two centuries ago, does not hold up in the real world of economic, political and social practices.

Woman's Nature: Realities versus Scientific Myths. As I said before, to be believed, scientific facts must fit the world-view of the times. Therefore, at times of tension and upheaval, such as the last two decades, some researchers always try to "prove" that differences in the political, social, and economic status of women and men, blacks and whites, or poor people and rich people, are inevitable because they are the results of people's inborn qualities and traits. Such scientists have tried to "prove" that blacks are innately less intelligent than whites, or that women are innately weaker, more nurturing, less good at math than men. If, for the purposes of this discussion, we focus on sex differences, it is clear that the ideology of woman's nature can differ drastically from the realities of women's lives and indeed be antithetical to them. In fact, the ideology functions, at least in part, to obscure the ways women live and to make people look away from the realities or ask misleading questions about them. So, for example, the ideology that labels women as the natural reproducers of the species, and men as producers of goods, has not been used to exempt women from also producing goods and services, but to shunt us out of higher paying jobs, the professions, and other kinds of work that require continuity and provide a measure of power over one's own and, at times, other people's lives. Most women who work for pay do so in job categories, such as secretary or nurse, which often involve a great deal of concealed responsibility, but are underpaid. This is one reason why insisting on equal pay *within* job categories cannot remedy women's economic disadvantage. Women will continue to be underpaid as long as women's jobs are less well paid than men's jobs and as long as access to traditional men's jobs is limited by social pressures, career counseling, training and hiring practices, trade union policies, and various other subtle and not so subtle societal mechanisms, such as research that "proves" that girls are not as good as boys at spatial perception, mathematics and science. An entire range of discriminatory practices is justified by the claim that they follow from the limits that biology places on women's capacity to work. Though exceptions are made during wars and other emergencies, they are forgotten as soon as life resumes its normal course. Then women are expected to return to their subordinate roles, not because the quality of their work during the emergencies has been inferior, but because these roles are seen as natural.

A few years ago, a number of women employees in the American chemical and automotive industries were actually forced to choose between working at relatively well-paying jobs that had previously been done by men or remaining fertile. In one instance, five women were required to submit to sterilization *by hysterectomy* in order to avoid being transferred from work in the lead pigment department at the American Cyanamid plant in Willow Island, West Virginia to janitorial work at considerably lower wages and benefits (Stellman and Henifin 1982). Even though none of these women was pregnant or planning a pregnancy in the near future (indeed, the husband of one had had a vasectomy), they were considered "potentially pregnant" unless they could prove that they were sterile. This goes on despite the fact that exposure to lead can damage sperm as well as eggs and can affect the health of workers (male and female) as well as a "potential fetus." It is as though fertile women are at all times potential parents; men, never. But it is important to notice that this vicious choice is being forced only on women who have recently entered relatively well-paid, traditionally male jobs. Women whose work routinely involves reproductive hazards because it exposes them to chemical or radiation hazards, but who have traditionally female jobs such as nurses, X-ray technologists, laboratory technicians, cleaning women in surgical operating rooms, scientific laboratories or the chemical and biotechnology industries, beauticians, secretaries, workers in the ceramics industry, and domestic workers are not warned about the chemical or physical hazards of their work to their health or to that of a fetus, should they be pregnant. In other words, scientific knowledge about fetal susceptibility to noxious chemicals and radiation is used to keep women out of better paid job categories from which they had previously been excluded by discriminatory employment practices, but, in general, women (or, indeed, men) are not protected against health endangering work.

The ideology of woman's nature that is invoked at these times would have us believe that a woman's capacity to become pregnant leaves her always physically disabled by comparison with men. The scientific underpinnings for these ideas were elaborated in the nineteenth century by the white, university-educated, mainly upper class men who made up the bulk of the new professions of obstetrics and gynecology, biology, psychology, sociology and anthropology. These professionals used their theories of women's innate frailty to disqualify the girls and women of their own race and class who would have been competing with them for education and professional status. They also realized that they might lose the kinds of personal attention they were accustomed to get from mothers, wives, and sisters if women of their own class gained access to the professions. They did not invoke women's weakness when it came to poor women spending long hours working in the homes and factories belonging to members of the upper classes, nor against the ways black slave women were made to work on the plantations and in the homes of their masters and mistresses.

Nineteenth century biologists and physicians claimed that women's brains were smaller than men's and that women's ovaries and uteruses required much energy and rest in order to function properly. They "proved" that therefore young girls must be kept away from schools and colleges once they begin to menstruate and warned that without this kind of care women's uteruses and ovaries will shrivel and the human race die out. Yet again, this analysis was not carried over to poor women, who were not only required to work hard, but often were said to reproduce *too* much. Indeed, scientists interpreted the fact that poor women could work hard and yet bear many children as a sign that they were more animal-like and less highly evolved than upper class women.

During the past decade, feminists have uncovered this history. We have analyzed the self-serving theories and documented the absurdity of the claims as well as their class and race biases and their glaringly political intent (Hubbard and Lowe 1979; Lowe and Hubbard 1983; Bleier 1984; Fausto-Sterling 1985). But this kind of scientific mythmaking is not past history. Just as in the nineteenth century medical men and biologists fought women's political organizing for equality by claiming that our reproductive organs made us unfit for anything but childbearing and childrearing, just as Freud declared women to be intrinsically less stable, intellectually inventive and productive than men, so beginning in the 1970's, there has been a renaissance in sex differences research that has claimed to prove scientifically that women are innately better than men at home care and mothering while men are innately better fitted than women for the competitive life of the market place.

Questionable experimental results obtained with animals (primarily that prototypic human, the white laboratory rat) are treated as though they can be applied equally well to people. On this basis, some scientists are now claiming that the secretion of different amounts of so-called male hormones (androgens) by male and female fetuses produces life-long differences in women's and men's brains. They claim not only that these (unproved) differences in fetal hormone levels exist, but imply (without evidence) that they predispose women and men *as groups* to exhibit innate differences in our abilities to localize objects in space, in our verbal and mathematical aptitudes, in aggressiveness and competitiveness, nurturing ability, and so on (Money and Ehrhardt 1972; Goy and McEwen 1980; *Science* 1981, 1263-1324). Sociobiologists claim that some of the sex differences in social behavior that exist in Western, capitalist societies (such as, aggressiveness, competitiveness, and dominance among men; coyness, nurturance, and submissiveness among women) are human universals that have existed in all times and cultures. Because these traits are said to be ever-present, sociobiologists deduce that they must have evolved through Darwinian natural selection and are now part of our genetic inheritance (Wilson 1975).

Sociobiologists have tried to prove that women's disproportionate contributions to child- and homecare are biologically programmed because women have a greater biological "investment" in our children than men have. They offer the following rationale: an organism's biological fitness, in the Darwinian sense, depends on producing the greatest possible number of offspring, who themselves survive long enough to reproduce, because this is what determines the frequency with which an individual's genes will be represented in successive generations. Following this logic a step further, sociobiologists argue that women and men must adopt basically different strategies to maximize opportunities to spread our genes into future generations. The calculus goes as follows: Eggs are larger than sperm and women can produce many fewer of them than men can sperm. Therefore each egg that develops into a child represents a much larger fraction of the total number of children a woman can produce, hence of her "reproductive fitness," than a sperm that becomes a child does of a man's "fitness." In addition, women "invest" the nine months of pregnancy in each child. Women must therefore be more careful than men to acquire well-endowed sex partners who will be good providers to make sure that their few investments (read, children) mature. Thus, from seemingly innocent biological asymmetries between sperm and eggs flow such major social consequences as female fidelity, male promiscuity, women's disproportional contribution to caring for home and children, and the unequal distribution of labor by sex. As sociobiologist, David Barash, says, "mother nature is sexist," so don't blame her human sons (Dawkins 1976; Barash 1979, esp. 46-90).

In devising these explanations, sociobiologists ignore the fact that human societies do not operate with a few superstuds; nor do stronger or more powerful men as a rule have more children than weaker ones. Men, in theory, could have many more children than women can, but in most societies equal numbers of men and women engage in producing children, though not in caring for them. These kinds of absurdities are useful to people who have a stake in maintaining present inequalities. They mystify procreation, yet have a superficial ring of plausibility and thus offer naturalistic justifications for discriminatory practices.

As the new scholarship on women has grown, a few anthropologists and biologists have tried to mitigate the male bias that underlies these kinds of theories by describing how females contribute to social life and species survival in important ways that are overlooked by scientists who think of females only in relation to reproduction and look to males for everything else (Lancaster 1975; Hrdy 1981, 1986; Kevles 1986). But, unless scientists challenge the basic premises that underlie the standard, male-centered descriptions and analyses, such revisions do not offer radically different formulations and insights. (For examples of more fundamental criticisms of evolutionary thinking and sociobiology, see Lowe and Hubbard 1979; Hubbard 1982; Lewontin, Rose and Kamin 1984).

Subjectivity and Objectivity. I want to come back to Paulo Freire, who says: "Reality is never just simply the objective datum, the concrete fact, but is also people's [and I would say, certain people's] perception of it." And he speaks of "the indispensable unity between subjectivity and objectivity in the act of knowing" (Freire 1985, 51).

The recognition of this "indispensable unity" is what feminist methodology is about. It is especially necessary for a feminist methodology in science because the scientific method rests on a particular definition of objectivity, that we feminists must call into question. Feminists and others who draw attention to the devices that the dominant group has used to deny other people access to power—be it political power or the power to make facts—have come to understand how that definition of objectivity functions in the processes of exclusion I discussed at the beginning.

Natural scientists attain their objectivity by looking upon nature (including other people) in small chunks and as isolated objects. They usually deny, or at least do not acknowledge, their relationship to the "objects" they study. In other words, natural scientists describe their activities as though they existed in a vacuum. The way language is used in scientific writing reinforces this illusion because it implicitly denies the relevance of time, place, social context, authorship, and personal responsibility. When I report a discovery, I do not write, "One sunny Monday after a restful weekend, I came into the laboratory, set up my experiment and shortly noticed that . . ." No; proper style dictates, "It has been observed that . . ." This removes relevance of time and place, and implies that the observation did not originate in the head of a human observer, specifically my head, but out there in the world. By deleting the scientist-agent as well as her or his participation as observer, people are left with the concept of science as a thing in itself, that truly reflects nature and that can be treated as though it were as real as, and indeed equivalent to, nature.

A blatant example of the kind of context-stripping that is commonly called objectivity is the way E. O. Wilson opens the final chapter of his *Sociobiology: The New Synthesis* (Wilson 1975, 547). He writes: "Let us now consider man in the free spirit of natural history, as though we were zoologists from another planet completing a catalog of social species on earth." That statement epitomizes the fallacy we need to get rid of. There is no "free spirit of natural history," only a set of descriptions put forward by the mostly white, educated, Euro-American men who have been practicing a particular kind of science during the past two hundred years. Nor do we have any idea what "zoologists from another planet" would have to say about "man" (which, I guess is supposed to mean "people") or about other "social species on earth," since that would depend on how these "zoologists" were used to living on their own planet and by what experiences they would therefore judge us. Feminists must insist that subjectivity and context cannot be stripped away, that they must be acknowledged if we want to use science as a way to understand nature and society and to use the knowledge we gain constructively.

For a different kind of example, take the economic concept of unemployment which in the United States has become "chronic unemployment" or even "the normal rate of unemployment." Such pseudo-objective phrases obscure a wealth of political and economic relationships which are subject to social action and change. By turning the activities of certain people who have the power to hire or not hire other people into depersonalized descriptions of economic fact, by turning activities of scientists into "factual" statements about nature or society, scientific language helps to mystify and intimidate the "lay public," those anonymous others, as well as scientists, and makes them feel powerless.

Another example of the absurdity of pretended objectivity, is a study that was described in the *New York Times* in which scientists suggested that they had identified eight characteristics in young children that were predictive of the likelihood that the children would later develop schizophrenia. The scientists were proposing a longitudinal study of such children as they grow up to assess the accuracy of these predictions. This is absurd because such experiments cannot be done. How do you find a "control" group for parents who have been told that their child exhibits five out of the eight characteristics, or worse yet, all eight characteristics thought to be predictive of schizophrenia?

Do you tell some parents that this is so although it isn't? Do you not tell some parents whose children have been so identified? Even if psychiatrists agreed on the diagnosis of schizophrenia—which they do not—this kind of research cannot be done objectively. And certainly cannot be done ethically, that is, without harming people.

The problem is that the context-stripping that worked reasonably well for the classical physics of falling bodies has become the model for how to do every kind of science. And this even though physicists since the beginning of this century have recognized that the experimenter is part of the experiment and influences its outcome. That insight produced Heisenberg's uncertainty principle in physics: the recognition that the operations the experimenter performs disturb the system so that it is impossible to specify simultaneously the position and momentum of atoms and elementary particles. So, how about standing the situation on its head and using the social sciences, where context stripping is clearly impossible, as a model and do all science in a way that acknowledges the experimenter as a self-conscious subject who lives, and does science, within the context in which the phenomena she or he observes occur? Anthropologists often try to take extensive field notes about a new culture as quickly as possible after they enter it, before they incorporate the perspective and expectations of that culture, because they realize that once they know the foreign culture well and feel at home in it, they will begin to take some of its most significant aspects for granted and stop seeing them. Yet they realize at the same time that they must also acknowledge the limitations their own personal and social backgrounds impose on the way they perceive the foreign society. Awareness of our subjectivity and context must be part of doing science because there is no way we can eliminate them. We come to the objects we study with our particular personal and social backgrounds and with inevitable interests. Once we acknowledge those, we can try to understand the world, so to speak, from inside instead of pretending to be objective outsiders looking in.

The social structure of the laboratory in which scientists work and the community and inter-personal relationships in which they live are also part of the subjective reality and context of doing science. Yet, we usually ignore them when we speak of a scientist's scientific work despite the fact that natural scientists work in highly organized social systems. Obviously, the sociology of laboratory life is structured by class, sex, and race, as is the rest of society. We saw before that to understand what goes on in the laboratory we must ask questions about who does what kinds of work. What does the lab chief—the person whose name appears on the stationery or on the door—contribute? How are decisions made about what work gets done and in what order? What role do women, whatever our class and race, or men of color and men from working class backgrounds play in this performance?

Note that women have played a very large role in the production of science—as wives, sisters, secretaries, technicians, and students of "great men"— though usually not as accredited scientists. One of our jobs as feminists must be to acknowledge that role. If feminists are to make a difference in the ways science is done and understood, we must not just try to become scientists who occupy the traditional structures, follow established patterns of behavior, and accept prevailing systems of explanation; we must understand and describe accurately the roles women have played all along in the process of making science. But we must also ask why certain ways of systematically interacting with nature and of using the knowledge so gained are acknowledged as science whereas others are not.

I am talking of the distinction between the laboratory and that other, quite differently structured, place of discovery and fact-making, the house-hold, where women use a different brand of botany, chemistry, and hygiene to work in our gardens, kitchens, nurseries, and sick rooms. Much of the knowledge women have acquired in those places is systematic and effective and has been handed on by word of mouth and in writing. But just as our society downgrades manual labor, it also downgrades knowledge that is produced in other than professional settings, however systematic it may be. It downgrades the orally transmitted knowledge and the unpaid observations, experimentation and teaching that happen in the household. Yet here is a wide range of systematic, empirical knowledge that has gone unnoticed and unvalidated (in fact, devalued and invalidated) by the institutions that catalog and describe, and thus define, what is to be called knowledge. Men's explorations of nature also began at home, but later were institutionalized and professionalized. Women's explorations have stayed close to home and their value has not been acknowledged.

What I am proposing is the opposite of the project the domestic science movement put forward at the turn of the century. That movement tried to make women's domestic work more "scientific" in the traditional sense of the word (Newman 1985, 156-191). I am suggesting that we acknowledge the scientific value of many of the facts and knowledge that women have accumulated and passed on in our homes and in volunteer organizations.

I doubt that women as gendered beings have something new or different to contribute to science, but women as political beings do. One of the most important things we must do is to insist on the political content of science and on its political role. The pretense that science is objective, apolitical and value-neutral is profoundly political because it obscures the political role that science and technology play in underwriting the existing distribution of power in society. Science and technology always operate in somebody's interest and serve someone or some group of people. To the extent that scientists are "neutral" that merely means that they support the existing distribution of interests and power

If we want to integrate feminist politics into our science, we must insist on the political nature and content of scientific work and of the way science is taught and otherwise communicated to the public. We must broaden the base of experience and knowledge on which scientists draw by making it possible for a wider range of people to do science, and to do it in different ways. We must also provide kinds of understanding that are useful and useable by a broad range of people. For this, science would have to be different from the way it is now. The important questions would have to be generated by a different social process. A wider range of people would have to have access to making scientific facts and to understanding and using them. Also, the process of validation would have to be under more public scrutiny, so that research topics and facts that benefit only a small elite while oppressing large segments of the population would not be acceptable.

Our present science, which supposedly exists to explain nature and let us live more comfortably in it, has in fact mystified nature. As Virginia Woolf's Orlando says as she enters a department store elevator:

> The very fabric of life now . . . is magic. In the eighteenth century, we knew how everything was done; but here I rise through the air; I listen to voices in America; I see men flying—but how it's done, I can't even begin to wonder. (Woolf 1928, 300)

Other ways to do Science? The most concrete examples of a different kind of science that I can think of come from the women's health movement and the process by which the Boston Women's Health Book Collective's (1984) *The* New *Our Bodies, Ourselves* or the Federation of Feminist Women's Health Centers' (1981) *A New View of a Woman's Body* have been generated. These groups have consciously tried to involve a range of women in setting the agenda, as well as in asking and answering the relevant questions. But there is probably no single way in which to change present-day science, and there shouldn't be. After all, one of the problems with science, as it exists now, is that scientists narrowly circumscribe the allowed ways to learn about nature and reject deviations as deviance.

Of course it is difficult for feminists who, as women, are just gaining a toehold in science, to try to make fundamental changes in the ways scientists perceive science and do it. This is why many scientists who are feminists live double-lives and conform to the pretenses of an apolitical, value-free, meritocratic science in our working lives while living our politics elsewhere. Meanwhile, many of us who want to integrate our politics with our work, analyze and critique the standard science, but no longer do it. Here again, feminist health centers and counselling groups come to mind as efforts to integrate feminist inquiry and political praxis. It would be important for feminists, who are trying to reconceptualize reality and reorganize knowledge and its uses in areas other than health, to create environments ("outstitutes") in which we can work together and communicate with other individuals and groups, so that people with different backgrounds and agendas can exchange questions, answers, and expertise.

NOTES

A different version of this paper will be published under the title, "Some Thoughts About the Masculinity of the Natural Sciences," in Mary Gergen, ed. *Feminist Knowledge Structures*. New York: New York University Press, in press.

1. Recently some physicists have hypothesized that a pound of feathers falls more *rapidly* than a pound of rocks—an even more counterintuitive "fact" than what I learned in high school physics.

REFERENCES

Barash, D. 1979. *The whispering within*. New York: Harper & Row.

Bleier, R. 1984. *Science and gender*. New York: Pergamon.

Boston Women's Health Book Collective. 1984. *The* new *our bodies, ourselves*. New York: Simon and Schuster.

Dawkins, R. 1976. *The selfish gene*. New York: Oxford University Press.

Fausto-Sterling, A. 1985. *Myths of gender*. New York: Basic Books.

Federation of Feminist Women's Health Centers. 1981. *A new view of a woman's body*. New York: Simon and Schuster.

Freire, P. 1985. *The politics of education*. South Hadley, MA: Bergin and Garvey.

Goy, R.W. and B.S. McEwen. 1980. *Sexual differentiation of the brain*. Cambridge, MA: M.I.T. Press.

Hrdy, S.B. 1981. *The woman that never evolved*. Cambridge, MA: Harvard University Press.

———. 1986. Empathy, polyandry, and the myth of the coy female. In *Feminist approaches to science*, ed. R. Bleier, 119-146. New York: Pergamon.

Hubbard, R. 1982. Have only men evolved? In *Biological woman—The convenient myth*, ed. R. Hubbard, M.S. Henifin and B. Fried, 17-46. Cambridge, MA: Schenkman.

Hubbard, R. and M. Lowe, eds. 1979. *Genes and gender II: Pitfalls in research on sex and gender*. Staten Island, NY: Gordian Press.

Kevles, B. 1986. *Females of the species*. Cambridge, MA: Harvard University Press.

Lancaster, J.B. 1975. *Primate behavior and the emergence of human culture*. New York: Holt, Rinehart and Winston.

Lewontin, R.C., S. Rose and L.J. Kamin. 1984. *Not in our genes*. New York: Pantheon.

Lowe, M. and R. Hubbard. 1979. Sociobiology and biosociology: Can science prove the biological basis of sex differences in behavior? In *Genes and gender II: Pitfalls in research on sex and gender*, ed. R. Hubbard and M. Lowe, 91-112. Staten Island, NY: Gordian Press.

Lowe, M. and R. Hubbard, eds. 1983. *Woman's nature: Rationalizations of inequality*. New York: Pergamon.

Money, J. and A.A. Ehrhardt. 1972. *Man & woman, boy & girl*. Baltimore: Johns Hopkins University Press.

Newman, L.M., ed. 1985. *Men's ideas/Women's realities: Popular science, 1870-1915*. New York: Pergamon.

Science. 1981. *211*, pp. 1263-1324.

Stellman, J.M. and M.S. Henifin. 1982. No fertile women need apply: Employment discrimination and reproductive hazards in the workplace. In *Biological woman—The convenient myth*, ed. R. Hubbard, M.S. Henifin and B. Fried, 117-145. Cambridge, MA: Schenkman.

Wilson, E.O. 1975. *Sociobiology: The new synthesis*. Cambridge, MA: Harvard University Press.

Woolf, V. 1928. *Orlando*. New York: Harcourt Brace Jovanovich; Harvest Paperback Edition.

Modeling the Gender Politics in Science

ELIZABETH POTTER

Feminist science scholars need models of science that allow feminist accounts, not only of the inception and reception of scientific theories, but of their content as well. I argue that a "Network Model," properly modified, makes clear theoretically how *race, sex and class considerations can influence the content of scientific theories. The adoption of the "corpuscular philosophy" by Robert Boyle and other Puritan scientists during the English Civil War offers us a good case on which to test such a model. According to these men, the minute corpuscles constituting the physical world are dead, not alive; passive, not active. I argue that they chose the principle that matter is passive in part because its contrary, the principle that matter is alive and self-moving, had a radical social meaning and use to the women and men working for progressive change in mid-seventeenth century England.*

Whenever she tells a story, the storyteller has a story model, one that outlines how a good story should go. When Marie Boas told the story of Robert Boyle's contributions to physics and chemistry in the mid-Seventeenth Century, she put her account of Boyle's life in a chapter separate from chapters describing his scientific achievements (Boas 1958). In her biography we discover that Boyle (1626-1691) was the youngest son of the fabulously wealthy, unscrupulous Earl of Cork. Old Cork was killed defending his holdings in the Irish rebellion of 1643 and his sons spent most of the Interregnum recovering their property in both England and Ireland. The young Robert was aided in this endeavor by the Parliamentary connections of his sister, Lady Katherine Ranelagh. We learn that Boyle never married and that, although he lived for many years in Oxford, he lived the last twenty years of his life in London with his sister, Katherine. He died a very rich man. As we turn to the chapters delineating his work in science, we find nothing to indicate that Boyle's class and sex or the meanings given to them by the turmoil of the Civil Wars, influenced his discoveries or his scientific theories in any way. But we do learn of a clear demarcation between the budding science of chemistry and non-scientific activities like alchemy, and that it was really chemistry that Boyle was practicing despite his interest in alchemy, natural magic and the occult. His adoption of the "corpuscular philosophy"

Hypatia vol. 3, no. 1 (Spring 1988).

in the mid-1650's followed upon his growing dissatisfaction with Aristotelianism and his search for a successor science. The story model that Boas' account follows is a familiar one. She adheres closely to the demarcation between science and non-science as well as to the distinction between an internal and an external account of science. Her account of Boyle's scientific achievements is an internal one, a history of scientific ideas making no reference to the political and social considerations that might have influenced Boyle's theory construction.

Clearly the philosophy of science Boas adopts will not allow a feminist story about science to be told. For one thing, traditional philosophy of science proscribes any feminist critique of Boyle's corpuscular theory as androcentric. On the traditional model of science, bad science is the result of social or personal influences infecting scientific procedures, while good science emerges from scrupulous adherence to scientific methodology. It follows that a rational reconstruction by historians and philosophers of science according to some favored logic of science is a sufficient explanation of good science. Thus, the feminist charge that Boyle's corpuscular philosophy rested on a sexist physical principle comes down to the claim that his theory is bad science because it was influenced by gender considerations. On Boas' story model, feminist scholars of science cannot argue both that Boyle was a good scientist who worked out a good theory *and* that gender politics influenced not only the inception and reception of his theory, but its content as well. But feminist scholars, along with other post-Kuhnian historians, sociologists and philosophers of science, have come to feel that considerations of race, sex and class may well belong in any adequate account of scientific work. Therefore, we need a way to understand science that will allow us to show how these considerations influence the way scientific theories are formulated.

The work of Mary Hesse, W.V.O. Quine and others offers a point of departure for understanding natural science in a way that will be useful for the new feminist accounts of science. In the following section, I will tell a feminist story about the adoption in England of the "corpuscular philosophy." The story illustrates the way in which a principle central to physical theory can be constrained by gender politics. Robert Boyle and other Puritan scientists who took up and developed the early atomic theory argued that the minute corpuscles constituting the physical world are dead, not alive; passive, not active. They chose the principle that matter is passive in part because its contrary, the principle that matter is alive and self-moving, had a social meaning and utility to the women and men working for change in mid-seventeenth century England. Because it collapses the distinction between an "internal" and an "external" history of science, my story can show how concern about gender influenced the formulation of a scientific theory. In the final section, I will show how what Mary Hesse calls a "Network Model" of scientific theories makes it clear *how* race, sex and class considerations can influence the content of scientific theories.

GENDER POLITICS IN THE MECHANICAL PHILOSOPHY

Christopher Hill sees two revolutions in the history of mid-seventeenth century England. The Puritan revolution successfully abolished feudal tenures and arbitrary taxes, and established the sovereignty of Parliament and common law. But Hill properly discerns

> another revolution which never happened, though from time to time it threatened. This might have established communal property, a far wider democracy in political and legal institutions, might have disestablished the state church and rejected the protestant ethic. (Hill 1982, 15)

Hill neglects to mention that it might also have established some degree of sexual equality and greater freedom for women, a measure of relief from the rule that women must be chaste, silent and obedient.

The radicals who worked for the failed revolution held a natural philosophy that grew out of certain theological heresies. These heresies in turn derived from the natural magic tradition whose sources included works attributed to Hermes Trismegistus, believed by Renaissance thinkers to have been an Egyptian priest at the time of Moses. Hylozooism, the principle that all matter is alive, was central to Hermeticism. We find in the *Corpus Hermiticum*, for example, that "All that is in the world, without exception, is in movement, and that which is in movement is also in life. Contemplate then the beautiful arrangement of the world and see that it is alive, and that all matter is full of life" (Yates 1964, 31 and 34).

Paracelsus, the sixteenth century physician, clearly conjoins the natural magic tradition with political dissent. "Magic," he said, "has power to experience and fathom things which are inaccessible to human reason." As a champion of the poor and oppressed, he held that law aids only the rich, not the poor; though he charged the rich high fees, he cured poor people for free. He advocated a reformation of religion and of society, including a redistribution of wealth and was nearly killed for supporting the peasants in the great 1525 uprising (Easlea 1980, 102).

The association between the natural magic tradition and political rebellion became clearer as the turmoils of the Seventeenth Century began. In 1600, Campanella was tortured by the Inquisition for rebelling against Spanish rule in Naples in order to set up a "universal republic." Campanella describes the republic in his *City of the Sun*, published in 1623 while he was imprisoned: it is led by a Hermetic magician and characterized not only by eugenics but also by communal ownership of property. Campanella imagines an ideal human community living together in love which mirrors his view of nature:

the basic constituents of both are credited with life and consciousness. Bodies "enjoy being together and cherish their reciprocal contact with one another," so much so that they abhor any vacuum that would destroy their "community" (Easlea 1980, 105).

In England, the pre-eminent Digger Gerrard Winstanley offers a clear case of the sectarian debt to the natural magic tradition. He held a kind of materialist pantheism which identified God with the created world and so placed the spirit of life and cause of motion within terrestrial and celestial bodies themselves:

> To know the secrets of nature is to know the works of God. . . . And indeed if you would know spiritual things, it is to know how the spirit or power of wisdom and life, causing motion or growth, dwells within and governs both the several bodies of the stars and planets in the heavens above; and the several bodies of the earth below, as grass, plants, fishes, beasts, birds and mankind. (Quoted by Hill 1982, 142)

These beliefs about nature harmonized perfectly with a revolutionary commitment to human equality.

Sectaries like Winstanley, however, were not the only ones who adopted views derived from the natural magic tradition. Under the influence of Jan Comenius and his disciple, Samuel Hartlib, Robert Boyle and many other mainstream Puritan reformers also adopted such views. Peter Rattansi (1968) observes that Boyle mentions in a letter that he was studying nature aided by "the glosses of Aristotle, Epicurus, Paracelsus, Harvey, Helmont, and other learned expositors of that instructive volume." He also says that "he was weaned away from Aristotelian principles partly by the opposition of Telesio, Campanella ('and his ingenious epitomist Comenius'), Bacon, Gassendi and Descartes among others." As Rattansi notes, ". . . many of the problems that obsessively recur in [Boyle's] work are of the sort that were of central importance for Hermeticism: the curative power of amulets and weapon-salves; stellar virtues; the Alkahest; and transmutation. No doubt, they are all explained in impeccably mechanical terms . . ., but their importance for him becomes less puzzling in the context of his earlier commitments" to hermeticism, alchemy and natural magic (Rattansi 1968, 131 and 139). And Boyle was typical of a group of moderate reformers who found in the natural magic tradition a promising alternative to the now defunct Aristotelian dogmas, all of whom played recognized roles in developing modern science: John Evelyn, Seth Ward, John Wilkins, John Wallis, William Petty, Walter Charleton and Christopher Wren (Rattansi 1963 and 1968).

In the late 1640's and early 1650's, these men repudiated the natural magic tradition and turned to atomism, to the corpuscular theory which Descartes and Gassendi formulated upon the basis of a revival of Epicurus and

Democritus. The traditional view that their conversion was brought about by recognition of the overwhelming superiority of the corpuscular theory avoids completely the social and political issues at stake. If we collapse the distinction between an "internal" and an "external" account of the conversion, we can see that the momentous shift to atomism was overdetermined, influenced by both "internal" and "external" considerations.

Boyle and his compatriots shared a widely held conviction that the natural order on one hand and the moral and social order on the other are mutually reflective. We see this clearly in the use made by sectaries of their natural philosophy to support their social philosophy. Winstanley, for example, urges that men can know God's will for themselves by looking at the world around them and do not need priests and bishops as intermediaries to tell them God's will because *all* of nature is full of God or Reason. Not only tithes, but the entire institution of a state church should be abolished. The depth of this challenge would have been clear to Charles I, whose famous aphorism, "No Bishop, no King," records his conviction that the monarchy and the state church were interdependent institutions.

For all their interest in the natural magic tradition, Boyle and the other moderate reformers wished to distinguish their views as sharply as possible from those of the sectaries whose class interests they opposed. This was tricky business. Simply adopting the corpuscular philosophy was insufficient to make the distinction since one of the most important principles at issue was whether matter is alive, full of the vital spirit and so able to *guide* and *move* itself. As long as God or the World Soul inhabits even the atoms, it inhabits all men and allows them to know all that they need to of spiritual matters without priestly mediation. Refutation of the sectaries' natural and social philosophies demands, then, at least that matter be passive, dead, brute or inert. Thus Boyle, in one of his first statements of the mechanical view, is very concerned to argue against hylozooism:

> . . . methinks we may, without absurdity, conceive, that God, . . . having resolved before the creation, to make such a world as this of ours, did divide . . . that matter, which he had provided, into an innumerable multitude of very variously figured corpuscles, and both connected those particles into such textures or particular bodies, and placed them in such situations, and put them into such motions, that by the assistance of his ordinary preserving concourse, the phenomena, which he intended should appear in the universe, must as orderly follow, and be exhibited by the bodies necessarily acting according to those impressions or laws, though they understand them not at all, as if each of those creatures had a design of self-preservation, and were furnished

> with knowledge and industry to prosecute it; and as if there were diffused through the universe an intelligent being, watchful over the publick good of it, and careful to administer all things wisely for the good of the particular parts of it, but so far forth as is consistent with the good of the whole. . . . (Quoted in Jacob 1972, 18)

Although atoms behave as if they consciously and continually follow the dictates of a world soul, in fact they are brute bits of dead matter, put in motion by God.

The adoption of the principle that matter is dead thus had clear implications for the class struggle in England during the Civil War. As Margaret and James Jacob point out, the mechanical philosophers produced a theory which "'outlawed' radicalism from the universe" (1980, 254). But the issue of whether matter is active or passive had implications for gender relations as well as for class relations. The Seventeenth Century saw the development of the ideal woman as a bourgeoise who was to marry and to stay at home minding the house; while married, she was to own no property. She had no voice in the Church or State (Thomas 1958, 43). Puritan marriage manuals continually reinforced the view that "the man when he loveth should remember his superiority" (Quoted by Thomas 1958, 43) and William Gough, in his popular manual *Of Domesticall Duties* of 1622 and 1634, flatly declared that "the extent of wive's subjection doth stretch itself very far, even to all things" (Stone 1977, 197).

The rise of sectarianism , with its view that God is in everything and everyone, threatened the sexual status quo. Keith Thomas points out that "[f]rom the very beginning the separatists laid great emphasis upon the spiritual equality of the two sexes. . . . And once admitted to the sect women had an equal share in church government. 'It followeth necessarily', wrote John Robinson, 'that one faithful man, yea, or woman either, may as truly and effectually loose and bind, both in heaven and earth, as all the ministers in the world'" (Thomas 1958, 44).

In England during the 1640's, women of the independent churches debated, voted, prophesied and even preached. The Leveller John Lilbourne remarked that "Every particular and individual man and woman that ever breathed in the world since [Adam and Eve] are and were by nature all equal and alike in power, dignity, authority and majesty, none of them having (by nature) any authority, dominion or magisterial power, one over . . . another."

Leveller women played their part in politics side by side with their husbands and brothers, and were among those who, for the first time in English history, acted politically *as women*. The story merits telling in some detail. In 1642, a group of women petitioned the House of Lords ". . . That Religion may be established, and present aid and assistance transported into Ireland for the reliefe of the distressed Protestants" (Quoted in Higgins 1973, 185). On February 1, 1642, about 400 working women, artisans, shop-girls and labourers gathered at parliament to get an answer to their petition. When the Duke of Lennox cried "away with these women, we were best have a Parliament of women," some of the women blocked his way and broke his staff of office (Higgins 1973, 185).

In August 1643, women, apparently about two or three thousand strong, again demonstrated at Parliament to demand an end to the war and to ". . . cry out for their slain and imprisoned husbands." The women grew restive and

> . . . From words they fell to blowes, [for after noon the women] came againe to the doore of the House at the upper staires head, as soone as they were past a part of the Trained Band that usually stood Centinell there, they thrust them downe by the head and shoulders, and would suffer none to come in or out of the Parliament house for two hours together . . . at last ten Troopers . . . came to passe by the women . . . whereupon they drew their Swords, and laid on some of them with their Swords flatwayes for a good space, which they [the women] regarded not, but enclosed them [the soldiers]; upon this they then cut them on the face and hands, and one woman lost her nose . . . and about an houre after the House was up, a Troope of horse came, and cudgelled such as staid, with their Kanes, and dispersed them. . . . (Higgins 1973, 193 and 195)

The women demonstrated at Parliament off and on throughout the 1640's, their activities reaching a pitch in 1649. Their petition of April, 1649, apparently signed by 10,000 women, demanded the release from prison of the four Leveller leaders, Lilbourne, Wedwyn, Prince and Overton, and complained of high food prices and taxes. Although the House of Commons refused at first to receive the petition, since women had no legal standing, on Wednesday, April 25, "The House of Commons sent out the Sergeant at Armes to the women to fetch in their Petition . . . but sent them this answer by the Serjeant at Armes to tell them by word of mouth . . . 'Mr Speaker (by direction of the House) hath commanded me to tell you, That the matter you petition about, is of an higher concernment then you understand, that the House gave an answer to your Husbands; and therefore that you are desired to goe home, and looke after your owne businesse, and meddle with your huswifery.'" But the women replied, "we are no whit satisfied with the answer you gave unto our husbands." This uppitiness prompted one newspaper to declare that the women petitioners were really demanding "let women weare the breeches" (Higgins 1973, 202-203, 213).

In April of 1649 the Leveller women decided to petition "considering, That we have an equal share and interest with men in the Commonwealth." By May the women were claiming that

> Since we are assured of our creation in the image of God, of an interest in Christ equal unto men, as also of a proportionate share in the freedoms of the commonwealth, we cannot but wonder and grieve that we should appear so despicable in your eyes as to be thought unworthy to petition or represent our grievances to this honourable House. Have we not an equal interest with the men of this nation in those liberties and securities contained in the *Petition of Right*, and other the good laws of the land? (Higgins 1973, 217)

These events were sure to generate alarm among men of the dominant classes. Sound Englishmen were quick to recognize in the new sectarian views—that all people are equal, and in particular, that women are at least spiritually equal to men since everyone alike has the spark of life within them—a serious threat to the sexual status quo. In 1646 Thomas Edwards vigorously opposed the granting of religious toleration on the grounds that men "should never have peace in their families more, or ever after have command of wives, children, servants" (Quoted by Stone 1977, 155).

Boyle, and others who later became mechanicists, were in and out of London in the late 1640's and would have read and heard accounts of the outrageous activities of women at Parliament. It was precisely during this period that Boyle wrote what we might call his "essays on women." These include the "Letter to Fidelia," dated London, December 2, 1647; the "Letter to Mrs. Dury," dated Stalbridge, April 15, 1647;[1] "The Duty of a Mother's Being a Nurse" and "The Martyrdom of Theodora." The Boyle revealed in these seldom-read texts eagerly resisted radical change in the social position of women or in the ideology of women. The "Letter to Mrs. Dury," for example, inveighs against Corsica's painting because painting "invites loose gallants to tempt them [the women who paint]," and may have been written in support of a bill then before Parliament forbidding women to wear makeup, while "The Duty of a Mother's Being a Nurse" argues that mothers should breastfeed their own children instead of hiring wet nurses for them. In these non-scientific works, most of them intended for a private audience, Boyle saw fit to expose a relentless concern that women occupy the domestic space being created for them by bourgeois liberal ideology. The exclusion of these concerns from his scientific writings may be more than a mere generic convention. Boyle was not the only mechanicist concerned with gender issues. Walter Charleton repudiated the natural magic tradition in favor of the mechanical philosophy in the early 1650's and produced a very important translation of Gassendi's work on atomism in 1654. But he was also the author of *The Ephesian Matron*, whose hero exclaims of women:

> You are the true *Hiena's*, that allure us with the fairness of your skins; and when folly hath brought us within your reach, you leap upon us and devour us. You are the traitors to Wisdom; the impediment to Industry . . . the clogs to virtue, and goads that drive us to all Vice, Impiety, and Ruine. You are the Fools Paradise, the Wiseman's Plague, and the grand Error of Nature. (Quoted by Easlea 1980, 242)

When Charleton represents women as "Errors of Nature," he suggests that a corrected nature would exclude the voice of women heard so often among the sectaries. At a metaphoric level, at least, he recognizes that his own and Boyle's gender politics were excluded from their scientific arguments by sheer force.

Boyle's famous lab assistant, Robert Hooke, on the other hand, plainly makes the connection. Body and Motion are, he claimed, the "Female" and "Male" of Nature, Body being the "Female or Mother Principle," "therefore rightly called by Aristotle and other Philosophers, *Materia*, Material Substance, or *Mater*; this being in itself without Life or Motion, without form, and void, and dark, a Power in it self wholly unactive, until it be, as it were, impregnated by the second Principle [Motion], which may represent the Pater, and may be call'd *Paternus*, *Spiritus*, or hylarchick Spirit. . . ." (Hooke 1682, 171-172).

Can we retheorize science in a way that allows us to show how gender concerns of the sort Boyle, Charleton and Hooke display get into their scientific work?

ARTICULATING THE MODEL

To many feminists, my account of Boyle's adoption of the hypothesis that matter is dead will not seem very heretical. It seems a likely story. That is because we have post-Kuhnian models of science very different from Marie Boas' and we recognize that feminist accounts of science require the collapse of traditional distinctions honored in Boas' account. But how we articulate a model is important. Some articulations will be useful for feminist scholars and others will not. We need to understand how making a scientific theory can be connected in myriad ways with other social and political activities. Traditionally, science studies concerned with social influences on science have followed Boas' model and have limited themselves to examining how scientists are influenced as they decide what questions to address. Just so, Ruth Doell argues that two sets of interests have prompted the development and use of in vitro fertilization (IVF) and the human genetic engineering it has made possible. On one hand, women in our society desire to bear children and want the option of bearing children when they are sterile. On the other, the scientists have a professional interest "in promoting the expansion of the research establishment and may, in this era of commercial application of biotechnology, have entrepreneurial interests as well" (Doell 1986, 11).

Although very important for our understanding of the relations between science and society, these studies leave undisturbed the assumption that the contents of scientific theories are, or should be, pure, uninfected by nonscientific considerations. Operating under that assumption, philosophers of science traditionally defended the purity of science against revelations such as those of Doell by distinguishing the "context of discovery" from the "context of justification." Thus motives leading a scientist to undertake research (the context of discovery) have no logical relation to the theory produced by the research (the context of justification). In the same way, the theory is logically independent of the uses and abuses to which it is put. The fundamental assumption here is that if political, economic, gender, racial or other nonscientific considerations influence the work of producing the scientific theory, the result will be bad science.

On this view, only bad scientific theory is infected by social and other illegitimate concerns; good science is objective, neutral science. Feminists have come to suspect, however, that even scientific theories that are "good" by all the standard criteria—e.g., they are simple, elegant, fruitful, internally consistent, externally coherent with the received paradigm, predictive and so on—are androcentric or sexist.[2] A useful model would then do two things: as well as allowing us to examine the masculinist interests influencing the inception and subsequent use of scientific theories, it would also allow us to show how those interests affect the content of scientific theories. Second, in doing that, it would show us how it is possible for a scientific theory "good" on all the traditional criteria to be androcentric or sexist. A really useful model of science will allow us to say not only that an otherwise unimpeachable theory was *initiated* for reasons that are inimical to the interests of women and has been *used* in ways that are bad for women, but also that and how an otherwise unimpeachable theory may *be* sexist or androcentric even when it is cognitively virtuous and is produced by men and women of good will. We begin to see that the dichotomy between "good science" and "bad science," like the dichotomy between internal and external accounts of science, hinders rather than helps a feminist understanding of science. If we are forced to say *either* that the early atomic theory was good because it was fruitful, coherent and covered the data, *or* that it was bad because it rested on a principle with androcentric, upper class social meanings, we are hindered in our attempt to understand the relations between the gender and the "technical" considerations that go into theory construction and the relations between, for example, women's activities at Parliament and Boyle's experimental activities at Lady Ranelagh's house.

For feminist purposes, a natural science model should allow theoretically not only for social changes in the structure of scientific institutions (e.g., more women scientists), but also for changes in the way theories are constructed. If a model doesn't allow analysis of real science, it can't show us where our problems lie and where changes must occur. But if a model describes actual scientific theories, not merely ideal ones, it should account for the possibility of feminist science in such a way as to allow it to be socially useful as well as cognitively virtuous and rigorous.

Finally, we would be suspicious of any model of the natural sciences that conflicted with much of the insightful work that feminist scholars have produced heretofore. For example, many feminist scholars of science have used Marxist grounds to argue that male scientists have an unconscious desire to dominate and control women and nature and that this desire leads to an unconscious bias toward theories that allow and foster domination and control. Using object relations theory, other feminist scholars have made very similar arguments (Keller 1983a and 1983b; Harding and Hintikka 1983). Thus, it would be well if our model could show *theoretically*, with some precision, how non-scientific assumptions, including particularly evaluative assumptions such as the desirability of domination and control, influence theoretical content.

A model such as Mary Hesse's Network Model will, if pushed far enough, be useful to feminists in the ways set out above (Hesse 1974). Hesse understands a scientific theory as a system of laws which has a very complex relation to nature. When the scientist establishes a law, Hesse argues, she classifies phenomena on the basis of resemblances among them. An astronomer might decide, for example, that this star is a red dwarf because it is similar to other red dwarfs; a neuroendocrinologist might try to determine whether the role of nitrogen 1 and nitrogen 3 of the imidazole ring of histidine in TRF is the same as their role in LRF. Any scientist is, then, constantly faced with decisions as to whether two things are similar enough to be classed together. But since phenomena are similar in some respects and different in others (none are identical; they differ at least in occuring at different times or occupying different locations), the question becomes, "Which respects are more important, the similar ones or the dissimilar ones?" When the data are all in—here observations of the respects in which phenomena do and do not resemble one another—decisions must be made about which data are significant. This is a fundamental case of "interpreting the data." Data alone, observations alone, do not determine a law or generalization; for example, we observe that whales swim in the water and so are like fish; but we also observe that they are live-bearing like mammals. Are they fish or mammals? Because similarity is not transitive, a decision must be made on grounds other than observed similarity. That is, *b* may resemble *a* and *b* may resemble *c*, but *a* and *c* do not therefore resemble one another; how, then, should we classify *b*? As an *a* or as a *c*? Since any decision here is underdetermined by the data, it has to be determined on other grounds.

One criterion at work in such a case is logical coherence throughout the system; however, we cannot claim that this criterion alone is sufficient to account for theory production. Scientists do not always decide between conflicting observations on the grounds that one generalization provides coherence with the greatest number of other generalizations. The problematic generalization may instead be the occasion to decide that most of the generalizations in the theory are wrong.

At this point, the mainstream philosophers who adopt a network model argue that scientists either do or should have recourse to cognitive virtues. Scientists hold or should hold certain assumptions about what constitute good systems of laws or "good theories." Just so, Quine has argued, the assumptions that good theories are "conservative" or are "simple" guide the scientist to make the decision that conserves most of what has been held true in the past, or the one that makes the system simpler (Quine 1978). Hesse refers to the virtues as "coherence conditions" and argues that they include assumptions such as the goodness of symmetry and of certain analogies, models and so on (Hesse 1974, 52). However, feminists, as critical science scholars, want to know, not what mainstream philosophers of science think scientists should do, but what scientists actually do. That is, we need to *look and see* what assumptions scientists actually hold to when they decide between conflicting generalizations. The feminist working hypothesis is that the assumptions guiding classificatory decisions may be androcentric or sexist.[3]

Unless we extend the Network Model by recognizing gender, class and race assumptions as "coherence conditions," the model will not be useful to feminists because it will not really collapse the internal/external distinction or the other distinctions hindering a feminist understanding of science. Symmetry, favored analogies and models, like the traditional cognitive virtues, are still "technical" considerations, suitable for an internalist account of scientific theory production. But if we crack the Network Model open by looking with feminist eyes at actual cases in order to discover coherence conditions, the model shows us how androcentric or sexist assumptions influence the content of scientific theories. At least for those cases in which a particular generalization is underdetermined by the data, as the generalization that matter is dead was when Boyle considered it, the decision as to which generalization to adopt must be based on other grounds than simple observation. Because all the generalizations in a system are logically interrelated, the adoption of one of a pair of conflicting generalizations such as "matter is dead" or "matter is alive" will have repercussions throughout the system and throughout related systems. The assumption of some cognitive virtue(s) can determine which repercussions are desirable, but so can the assumption of some other principle, for instance that male behavior is the norm, that male behavior is crucial to evolution, that hierarchies are functional, that hierarchical models are better than nonhierarchical ones, or, as I have argued in Boyle's case, that women should be kept in a secondary social position. The suggestion here is that feminist studies of theory construction in the sciences should look carefully at the constraints affecting the choices scientists make between conflicting generalizations. On a network model each generalization in the system is—at any given time, though not at all times—corrigible, so there is nothing theoretically to prevent us from discovering that even the most innocent choice is constrained ultimately by an androcentric or sexist assumption.

The flexibility of any system of generalizations means that choices among generalizations can be made that allow at once some degree of empirical adequacy, of coherence, of fruitfulness, simplicity, faithfulness to preferred analogies or models *and* the maintenance of androcentric or sexist assumptions. Thus, the model makes it clear that even good scientific theories, by all the traditional criteria, can be androcentric or sexist in the sense that a sexist or androcentric assumption constrains the distribution of truth values throughout the system. Just so, the early atomic theory, as worked out by Gassendi, Descartes, Boyle and the others, had a reasonable degree of empirical adequacy, was fairly coherent, was certainly fruitful and so on. But one of its fundamental principles had a sexist as well as a classist social meaning. Since the hylozooic principle meant, if not full equality for women, then at least spiritual equality and certain corresponding liberties, for example, to speak and to act as their own consciences directed them, any passivity principle meant the loss of equality and of the corresponding liberties.

The flexibility of theoretical systems also allows the possibility of new and different theory constructions. We might speculate about how an atomic theory based on the hylozooic principle would have developed just as we might speculate about the current possibilities for feminist ways of doing science. Theories could be constructed that base classificatory decisions on feminist assumptions instead of antifeminist ones. The maintenance of a feminist assumption, such as the assumption that women were significant in human evolution or that nonhierarchical, organismic models are better than hierarchical ones, or that women and men should be allowed equal social positions, would have repercussions throughout the classificatory system and would still allow, just as nonfeminist assumptions have, some degree of empirical adequacy, of coherence, of fruitfulness, simplicity, faithfulness to preferred analogies or models, and so on.[4] In her book, *The Principles of the Most Ancient and Modern Philosophy*, Boyle's contemporary, Anne Conway (1982), began work on an atomic theory based on the principle that all matter is alive and conscious.[5] The full story of her work remains to be told, but a network model shows theoretically that such a corpuscular philosophy could have exhibited, just as Boyle's corpuscular philosophy did, a reasonable degree of empirical adequacy, coherence and fruitfulness, and could still have been politically progressive for women.

Notes

I am indebted to Kathryn Pyne Addelson, Sandra Harding and Caroline Whitbeck for helpful comments and suggestions on earlier drafts of this essay. I have also benefited from the comments and questions of college and university audiences in the Philadelphia area.

1. I include these letters because they are clearly set pieces written to a fictitious character (Fidelia) or about a fictitious character (Corisca).

2. Here an "androcentric" hypothesis is one that rests on assumptions made only from a male point of view, while a "sexist" hypothesis is one that rests on the assumption that women are inferior to men.

3. David Bloor (1982) argues that coherence conditions should include any social interests that influence the choice a scientist makes between competing hypotheses. Although he never mentions gender considerations, he might include androcentric or sexist social interests as coherence conditions. But there is a problem with the notion of "social interest" here: a woman scientist might well adopt a sexist or androcentric coherence condition even when it is against her social interest to do so (Cf. also Longino 1983). On the model set out here, what Longino refers to as "background assumptions" determine scientists' classificatory decisions.

4. The sociobiological theories of Sara Hrdy (1981), Nancy Tanner (1981) and Adrienne Zihlman beautifully illustrate this point (Cf. Hrdy 1981, Tanner 1981, Tanner and Zihlman 1976, and Zihlman 1978).

5. Carolyn Merchant (1980) discusses Conway's work and argues that Leibniz' monadology was developed under its influence (Merchant 1979).

References

Bloor, David. 1982. Durkheim and Mauss revisited: Classification and the sociology of knowledge. *Studies in the History and Philosophy of Science* 13:267-297.

Boas, Marie. 1958. *Robert Boyle and seventeenth-century chemistry.* Millwood, NY: Kraus Reprint Co.

Conway, Anne. 1982. *The principles of the most ancient and modern philosophy.* The Hague: Martinus Nijhoff.

Doell, Ruth. 1986. Agency and value judgment: A perspective from biology. Forthcoming in *Critical issues in feminist inquiry*, ed. Joan E. Hartman and Ellen Messer-Davidow. Modern Language Association.

Easlea, Brian. 1980. *Witch hunting, magic and the new philosophy: An introduction to debates of the scientific revolution 1450-1750.* New Jersey: Humanities Press.

Harding, Sandra and Merrill Hintikka, eds. 1983. *Discovering reality.* Boston: D. Reidel.

Hesse, Mary. 1974. *The structure of scientific inference.* Berkeley and Los Angeles: University of California Press.

Hill, Christopher. 1982. *The world turned upside down*. New York: Penguin Books.

Higgins, Patricia. 1973. The reactions of women. In *Politics, religion and the English civil war*, ed. Brian Manning. New York: St. Martin's Press.

Hooke, Robert. 1682. A discourse on the nature of comets. In *The posthumous works of Robert Hooke*. 1705. Ed. R. Waller. London.

Hrdy, Sara Blaffer. 1981. *The woman that never evolved*. Cambridge, MA: Harvard University Press.

Jacob, J.R. 1972. *Journal of European Studies* 2:1-21.

Jacob, James R. and Margaret C. Jacob. 1980. The anglican origins of modern science: The metaphysical foundations of the whig constitution. *Isis* 71:251-267.

Jacob, Margaret C. 1976. *The newtonians and the English revolution, 1689-1720*. Ithica: Cornell University Press.

Keller, Evelyn. 1983a. Gender and science. In *Discovering reality*. See Harding and Hintikka 1983.

———. 1983b. Feminism and science. In *The signs reader: Women, gender and scholarship*, ed. Elizabeth Abel and Emily K. Abel. Chicago and London: University of Chicago Press.

Longino, Helen. 1983. Beyond bad science: Skeptical reflections on the value-freedom of scientific inquiry. *Science, Technology and Human Values* 8:7-17.

Longino, Helen and Ruth Doell. 1983. Body, bias and behavior: A comparative analysis of reasoning in two areas of biological science. *Signs* 9:206-227.

Merchant, Carolyn. 1979. The vitalism of Anne Conway: Its impact on Leibniz's conception of the monad. *Journal of the History of Philosophy* Vol. 7.

———. 1980. *The death of nature*. San Francisco: Harper and Row.

Quine, W.V.O. 1978. *The web of belief*. New York: Random House.

Rattansi, Peter. 1963. Paracelsus and the puritan revolution. *Ambix* 11:24-32.

———. 1968. The intellectual origins of the royal society. *Notes and Records of the Royal Society* 23:129-143.

Stone, Lawrence. 1977. *The family, sex and marriage in England 1500-1800*. New York: Harper and Row.

Tanner, Nancy. 1981. *On becoming human*. Cambridge: Cambridge University Press.

Tanner, Nancy and Adrienne Zihlman. 1976. Women in evolution. Part I. Innovation and selection in human origins. *Signs* 1:585-608.

Thomas, Keith. 1958. Women and the civil war sects. *Past and Present* 13:42-62.

Yates, Frances A. 1964. *Giordano Bruno and the hermetic tradition*. New York: Vintage.

Zihlman, Adrienne. 1978. Women in evolution. Part II. Subsistence and social organization among early hominids. *Signs* 4:4-20.

The Weaker Seed
The Sexist Bias of Reproductive Theory

NANCY TUANA

This history of reproductive theories from Aristotle to the preformationists provides an excellent illustration of the ways in which the gender/science system informs the process of scientific investigation. In this essay I examine the effects of the bias of woman's inferiority upon theories of human reproduction. I argue that the adherence to a belief in the inferiority of the female creative principle biased scientific perception of the nature of woman's role in human generation.

Woman as other, the second sex, the weaker vessel, a misbegotten man. We are painfully aware of the litany of centuries of theories condemning women to a subordinate status.

Scientists, like everyone, work within and through the worldview of their time. The theories they develop and the facts they accept must be coherent with this system of beliefs. Thus, it will come as no surprise that science has provided a biological explanation/justification of the image of woman as inferior. The history of embryological and reproductive theories provides a revealing case study of the effects of this presupposition of woman's inferiority upon the generation of scientific theories. From Aristotle to the reproductive theories of the 1700s we can trace a pattern of depreciation of the female principle in conception originating from the assumption of woman's biological inferiority.

COLD WOMEN AND BARREN WOMBS

The first systematic development of a scientific explanation of woman's inferiority is found in the writings of Aristotle (384-322 BCE). Although Aristotle certainly did not invent the idea of the greater perfection of the male sex, his writings provided a detailed and seemingly rational justification of this belief that was to be accepted for centuries.

Aristotle's biology is based on the central premise that heat is the fundamental principle in the perfection of animals. Heat serves to "concoct" matter, that is, to enable it to develop. The more heat an animal is able to generate, the more developed it will be. "[T]hat which has by nature a smaller

Hypatia vol. 3, no. 1 (Spring 1988).

portion of heat is weaker" (GA 726.b.33). On the Aristotelian model, if woman is to be viewed as biologically inferior to man, it must be possible to demonstrate that woman generates less heat than man.

The belief that woman is colder than man is a central premise of Aristotle's biology. He employs it to account for numerous alleged physiological and psychological differences between women and men, and to justify the perception of these differences as "defects." Because woman has less heat she will be smaller and weaker than man. "The male is larger and longer-lived than the female . . . the female is less muscular and less compactly jointed" (HA 538.a.23-538.b.8). Woman's defect in heat will result in her brain being smaller and less developed than man's. "Of all the animals, man's brain is much the largest and the moistest" (GA 784.a.2-3). And woman's inferior brain size in turn accounts for much of her defective nature. Woman is:

> . . . more jealous, more querulous, more apt to scold and to strike. She is, furthermore, more prone to despondency and less hopeful than the man, more void of shame, more false of speech, more deceptive, and of more retentive memory. (HA 608.b.10-12)

Aristotle's evidence of woman's defect in heat is tied to his theory of reproduction. According to this theory, semen is derived from blood. Aristotle labels the male ejaculate "semen" and notes that male semen does not resemble blood, but is white in color. He insists that the appearance of male semen has to be the result of heat. Since a substance is transformed by being concocted and heat is required for such concoction, the appearance of male semen must be the result of an infusion of heat that concentrates the potency of the blood and changes its appearance.

Aristotle's account of female semen rests firmly upon the assumption of woman's inferiority. Aristotle equates menses with male semen. Menstrual discharge "is analogous in females to the semen in males" (GA 727.a.3-4). His justification for this is that "semen begins to appear in males and to be emitted at the same time of life that the menstrual flow begins in females" and that "in the decline of life the generative power fails in the one sex and the menstrual discharge in the other" (GA 727.a.5-10). Since the menstrual flow, like the ability to emit semen, commences at puberty and ceases with old age, Aristotle correctly concludes that it has to be associated with reproduction. However, Aristotle incorrectly assumes that female semen, like male semen, will be visibly emitted. Because of this he identifies menses as female semen.

Aristotle's proof of woman's defect in heat should now be obvious. Female semen is abundant and resembles blood. Male semen is scarce and is quite unlike blood. Aristotle claims that these differences are accounted for by the fact that women are unable to "cook" their semen to the point of purity—

thus "proof" of their relative coldness. Lacking heat in comparison to man, woman's semen is not transformed and looks like blood. It is also more abundant because woman is unable to reduce it through the infusion of heat. "[I]t is necessary that the weaker animal also should have a residue greater in quantity and less concocted" (GA 726.b.30-31).

Aristotle further concludes that female semen is impotent. "[I]t is plain that the female does not contribute semen to the generation of the offspring. For if she had semen she would not have menstrual fluid; but, as it is, because she has the latter she has not the former" (GA 727.a.27-30). In other words, woman's menstrual fluid is the blood that would be turned into potent semen if she had sufficient heat to concoct it. Because of woman's deficient heat, she is unable to transform this blood and turn it into seed.

The untenability of this view rests on the equation of the menstrual fluids of women with the seminal fluids of men, for there are striking differences between menses and semen. The most obvious difference is that menses occur only once a month and are not associated with intercourse or genital stimulation, while male seminal fluid is emitted whenever there is sufficient stimulation for ejaculation. Given this, it would have been logical for Aristotle to have considered whether females have a fluid emitted during intercourse or other genital stimulation that was the analogue to male semen. In fact, such a view was popular enough that Aristotle thought it important to refute it. "Some think that the female contributes semen in coition because the pleasure she experiences is sometimes similar to that of the male, and also is attended by a liquid discharge" (GA 727.b.34-36).

Aristotle offers three arguments designed to prove this view false. In the first, he insists that this discharge does not occur in all women. It appears only "in those who are fair-skinned and of a feminine type generally, but not in those who are dark and of a masculine appearance" (GA 728.a.2-4). (He gave no explanation of how he arrived at this observation!) Since some women do not have this discharge, yet all can conceive, Aristotle concludes that it cannot constitute woman's seminal fluid.

The second argument against labelling these emissions seminal is that "the pleasure of intercourse is caused by touch in the same region of the female as of the male; and yet it is not from thence that this flow proceeds" (GA 728.a.31-33). Aristotle appears to be arguing that it would be incorrect to associate this discharge with male semen for it is unlike semen in that it does not issue from the place of stimulation. It should be obvious by now that these arguments are contrived. Aristotle is straining against observation in his attempts to support his theory. All women have lubricating fluids. These fluids can be produced in a woman by stimulating the vulva or vagina, thus causing them to issue from the place touched. Furthermore, a male can ejaculate from stimulation of areas other than his penis, in which case the flow proceeds from a region other than that touched.

The third argument reveals this same tension. Aristotle insists that arguments associating these lubricating fluids with female semen were fallacious because the female often conceives without the sensation of pleasure in intercourse (GA 727.b.6-10). Although not explicitly stated, Aristotle seems to be assuming that seminal fluids are those that are emitted when sufficient levels of sexual pleasure are experienced. Since women conceive without experiencing pleasure in intercourse, and thus without emitting such fluids, and, conversely, experience pleasure in intercourse and thereby emit such fluids without conceiving, there is reason to deny the association of these fluids with semen. The weakness of this argument is demonstrated by the obvious fact that a male frequently ejaculates during intercourse without this resulting in pregnancy. But Aristotle maintains that the fact that a woman can emit her lubricating fluids without conceiving is proof that they cannot be seminal. Thus one of the same reasons given for denying the association of this female liquid with semen will call into question the accuracy of associating the male ejaculate with semen.

Aristotle's thesis that women produce no semen had numerous ramifications upon his understanding of human creation. In a striking passage, Aristotle supports the view that the fetus is contained within the male and placed by him into the female. "By definition the male is that which is able to generate in another . . . the female is that which is able to generate in itself and out of which comes into being the offspring *previously existing in the generator*" (GA 716.a.20-23, emphasis added).

Careful attention to Aristotle's view reveals that this offspring should not be thought of as existing preformed in the body of the male. Rather the male imparts to the female the form of the fetus. The female role in reproduction is to provide the material, the menstrual blood, upon which the male semen imparts form. The metaphor Aristotle used to explain this process is very similar to the craftsperson metaphor of creation as found in Genesis 2.7, where the divine being gives form to unformed substance. Aristotle compares the process of conception to that of a carpenter carving out a bed. In conception the female provides the raw material, just as the tree provides the wood. But it is the male, similar to the carpenter, who determines the function of the object to be produced and gives it its form. The female body thus becomes the workplace and source of raw material out of which the male crafts a human life (GA 729.b.15-73.b.25). Just as the god of Genesis imposed form upon the dust of the earth and breathed life into it, the seed of the male imposes the form and nature of a human being onto the blood of a woman's womb.

A value underlying this metaphor is that the female role in generation, though necessary, is relatively unimportant. The female merely provides the raw material out of which a fetus will be formed. For Aristotle, the material cause involved the least degree of being or perfection of all the four causes:

> . . . the first efficient or moving cause, to which belong the definition and the form, is better and more divine in its nature than the material on which it works . . . for the first principle of movement, whereby that which comes into being is male, is better and more divine, and the female is matter. (GA 732.a.3-9)

Woman is thus relegated to the least significant of all the causes. Man, in providing the form and motion of the fetus, is far more responsible for its generation.

Aristotle attempts to reinforce his thesis of the excellence of the male contribution to generation by arguing that it, unlike the female contribution, involves nothing material. The assumption implicit here, of course, is that nonmaterial substance is superior to (has a higher degree of perfection than) material substance. Again using the analogy of the carpenter, Arstotle depicts male semen as a tool that moves upon the material within the womb and imparts to it form and motion, but does not become a part of the material of the fetus. According to Aristotle, when a male ejaculates that "which is emitted [is] the principle of soul" (GA 737.b.7-8). The matter that is ejaculated with it dissolves and evaporates and is not a part of the generation process. Aristotle associates the male contribution with "something divine." "While the body is from the female, it is the soul that is from the male, for the soul is the substance of a particular body" (GA 738.b.25-26). The male contribution thus gives form and soul to the offspring. Once again the female contribution is shown to be inferior—it lacks this principle of soul. "For the female is, as it were, a mutilated male, and the menstrual fluids are semen, only not pure; for there is only one thing they have not in them, the principle of soul" (GA 737.a.27-29).

Aristotle here echoes the words of Aeschylus' Apollo, "[t]he woman you call the mother of the child/is not the parent, just a nurse to the seed,/ the new-sown seed that grows and swells inside her./ The *man* is the source of life" (Aeschylus 1975, 666-669). On the Aristotelian metaphysic, the end of a thing is more truly present in that which provides the form and source of the movement, than in that which provides the matter. But it is the male that is the source of movement in generation, for he imparts movement to the semen, which in turn imparts movement to the menstrual fluid. This view is reflected in the fact that Aristotle attributes to the male the active role in conception, while insisting that the female plays only a passive role. "But the female, as female, is passive, and the male, as male, is active, and the principle of movement comes from him" (GA 729.b.13-14). In generation the female is acted upon, is merely the passive matter upon which the artist works. Thus, the true creative act in generation arises from the actions of the male. It is the male who is the parent in the full sense, for it is only he that imparts form and movement to the offspring. The most creative aspects of procreation are attributed exclusively to man, while the female principle in generation is discounted as playing only the passive and least significant role of nursing the sown seed.

Having thus documented woman's defect in heat and delineated its implications concerning woman's nature and role in reproduction, Aristotle still had to offer an explanation of the cause of this defect. Although Aristotle attempted such an account, this part of his theory would soon be overturned by the work of Galen who claimed, quite correctly, that Aristotle merely begged the question.

To account for the cause of woman's innate defect in heat, Aristotle insists that an embryo becomes female "when the first principle does not bear sway and cannot concoct the nourishment through lack of heat nor bring it into its proper form" (GA 766.a.17-21). This is not an explanation of the cause of the defect, but merely an assertion of it. Galen was quite right in attacking this portion of Aristotle's theory. However, the statement reveals two important biases. First, Aristotle believes that the "proper form" of a human is male. In this respect the female is not fully human. Aristotle offers no proof or explanation of his claim. It is an axiom resulting from his assumption of woman's inferiority.

The second bias, stemming from the first, is that a female embryo is caused by a deviation from nature. On the Aristotelian account, nature always aims to create the most perfectly formed being—a male given the previous bias. If there is some lack of generative heat or some adversity, the creation will not be perfect and a female results. Hence, woman is a misbegotten man, resulting from some defect in the heat of the generative process.

The evidence Aristotle offers in support of such claims demonstrates the extent to which his biases affected his science. According to him, "observed facts confirm what we have said. For more females are produced by the young and by those verging on old age than by those in the prime of life; in the former the heat is not yet perfect, in the latter it is failing" (GA 766.b.28-30). Careful observations would quickly refute such a claim.

Another contention that would not have stood careful testing was that the greater heat of the male fetus results in its developing more quickly than the female. Aristotle claims that the male fetus first moves in the womb about the fortieth day, while the female does not move until the ninetieth day (HA 583.b.3-5). This is a reasonable inference from his theory, but one that could be easily disconfirmed by carefully recording fetal movement in humans.

Furthermore, Aristotle's theory is full of inconsistencies. He insists that woman is a mutation, resulting as she does from some defect in heat. "The female is, as it were, a mutilated male" (GA 737.a.27-28). Yet he notes that males are more often born defective than females. It would seem logical here to conclude that such higher rates of mutations in males is due to a defect in heat, thus arguing against the premise of male physiological superiority. Aristotle, however, manages to turn the higher preponderance of birth defects in males into a mark of perfection! He insists that male defects are caused not by a defect in heat, but rather by man's superior heat. Twisting logic into knots, Aristotle insists that due to male's greater natural heat, the male fetus "moves about more than the female, and on account of moving is more liable to injury" (GA 775.a.6-8).

Aristotle also notices that females reach puberty more quickly than do males. On the basis of his earlier argument in which he claimed that accelerated development was a mark of superior heat, one would expect Aristotle to conclude that females have a greater amount of heat than males. Through yet another twist of logic, Aristotle insists that puberty rates are an additional proof of woman's defect in heat!

> For females are weaker and colder in nature, and we must look upon the female character as being a sort of natural deficiency. accordingly while it is within the mother it develops slowly because of its coldness (for development is concoction, and it is heat that concocts, and what is hotter is easily concocted); but after birth it quickly arrives at maturity and old age on account of its weakness, for all inferior things come sooner to their perfection, and as this is true of works of art so it is of what is formed by nature. (GA 775.a.13-21)

We can see from such inconsistencies in Aristotle's theory that the doctrine that the female sex was inferior to the male was not a premise to be proved or justified, but was rather an implicit belief underlying Aristotle's development of his biological theory and an axiom upon which he founded his theory of reproduction.

The Phallic Vagina

Aristotle's view of the primacy of the male principle of creation was consistent both with the metaphysical views of his time as well as with other scientific theories of generation. His embryological theory, however, was extreme in holding that only the male contributed seed. Many classical theorists including Anaxagoras, Empedocles, Hippocrates and Parmenides held that the fetus was the result of a combination of male and female semen. Although such theorists gave woman a role in the creation of the form as well as the material of the fetus, they uniformly held that woman's contribution was weaker than that of man.

This contrasting view was supported by Galen (130?-?200). Although Galen supported much of the biology of Aristotle, he offered certain "corrections" and developments. He accepted Aristotle's basic position that a defect in heat caused woman's inferiority. "The female is less perfect than the male for one, principle reason—because she is colder" (14.II.295). However he argued that Aristotle overlooked one of the most obvious indications of woman's inferiority—her genitals. "Aristotle was right in thinking the female less perfect than the male; he certainly did not, however follow out his argument to its conclusion, but, as it seems to me, left out the main head of it, so to speak" (14.II.295).

Galen insists that upon dissection of the genitals of the two sexes, one will realize that all the parts that men have, women have too. He claims that the only difference that exists between the genitals of the two sexes is that in women the various parts of them are within the body, whereas in men they are outside the body. "Consider first whichever ones you please, turn outward the woman's, turn inward, so to speak, and fold double the man's, and you will find them the same in both in every respect" (14.II.297).

Galen's explanation of the source of this difference will not be a surprise. Woman's genitals "were formed within her when she was still a fetus, but could not because of the defect in the heat emerge and project on the outside" (14.II.299). The internal location of woman's genitals is thus the "proof" that Aristotle overlooked. Once Galen has equated male and female genitals, he insists that the internal location of woman's genitals can only be explained by an arrested development. Fully concocted genitals, the "true" form of genitals, are, of course, the penis and testicles. Woman remains, so to speak, half-baked.

The history of anatomy attests to the power of Galen's analysis of woman's genitals. For centuries, anatomical drawings of women's internal genitalia would bear an uncanny resemblance to man's external genitalia. (See illustration I.)

Having thus "demonstrated" the imperfection of woman's genitals, Galen predictably concludes that woman's semen is similarly imperfect. "[T]he female must have smaller, less perfect testes, and the semen generated in them must be scantier, colder, and wetter (for these things too follow of necessity from the deficient heat)" (14.II.301). Galen thus deviates from the Aristotelian position that woman contributes no seed in generation. Woman contributes seed, but it is inferior to man's seed.

Galen now turned to the source of woman's defect in heat. You will remember that Aristotle offered no convincing account of the physiological source of this defect. Galen, aware of this omission, was concerned to address it. Galen accepted as true the popular Hippocratic view that male seed is produced in the right testis and female seed in the left testis (Hippocrates 1943, XXVII & XXIX). According to this theory, a girl would be produced when seed from the left testis (ovary) of the female combined with seed from the left testis of the male and implanted in the left side of the uterus, for the left side of the uterus was believed to provide nourishment that results

The twelfth Figure, of the Wombe.

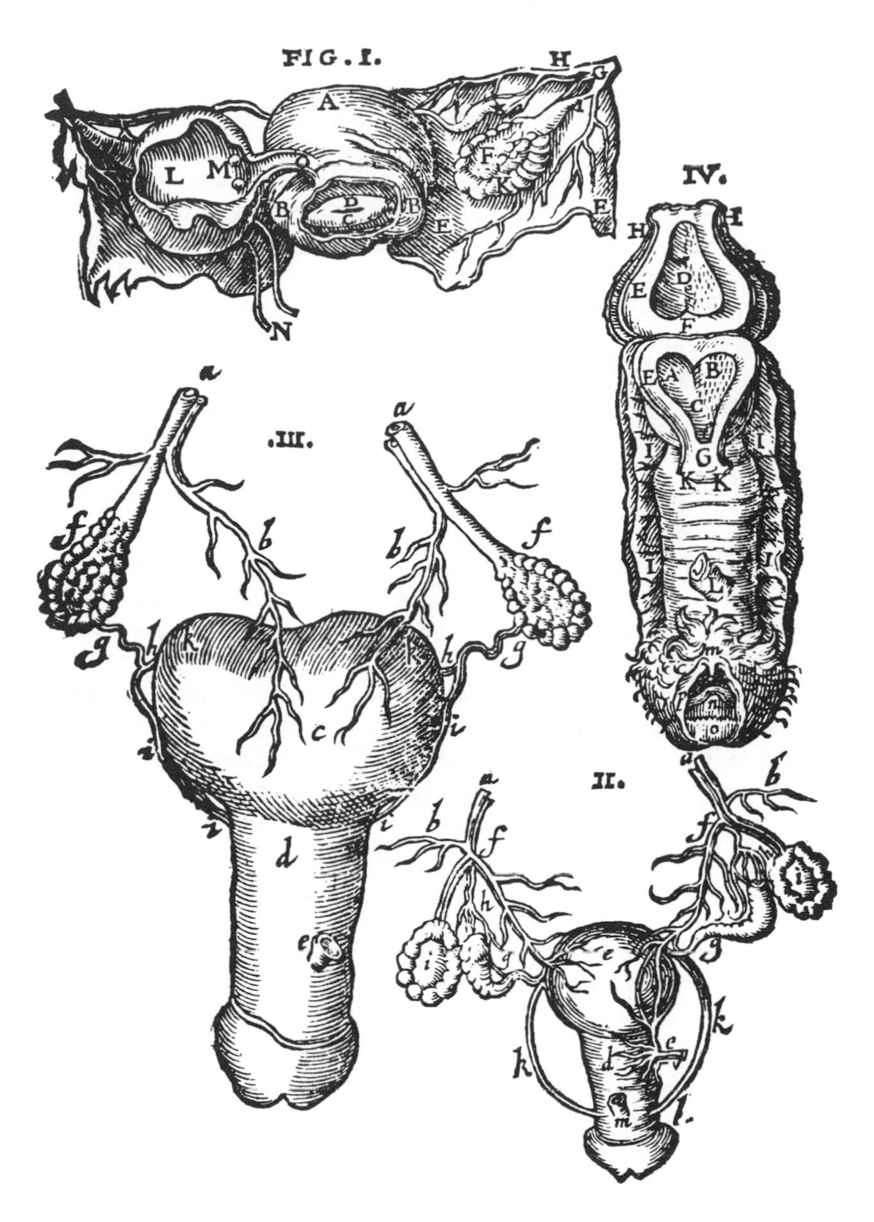

Illustration I. These illustrations are found in the writings of Ambroise Paré (1517?-1590). Notice the phallic representation of the female genitalia.

in female characteristics.

On the basis of this belief that the seed out of which a woman evolves originates from the left testes while male seed originates from the right, Galen is able to construct a biological explanation of the inferiority of female seed and thereby explain the physiological source of woman's defect in heat. Using very creative anatomy, Galen explains that the spermatic and ovarian artery and vein going to the right testis and right side of the uterus arise directly from the vena cava and the aorta below the level of the renal vessels and thus carry blood already cleansed by the kidneys. The corresponding veins and arteries of the left testis arise from the renal vessels going to the left kidney and thus carry uncleansed blood. (See illustration II.) We now know that the veins and arteries of both sides arise from the vena cava and aorta, but this mistake by Galen proved to be a very convenient explanation of the cause of woman's biological inferiority. Holding that they are fed from veins that pass to the kidneys, Galen explains that "the left testis in the male and the left uterus in the female receive blood still uncleansed, full of residues, watery and serous" (14.II.306). In other words, the blood going to the left side is impure. Because it is impure, such blood contains less heat, which will result in its producing imperfect seed. The male seed, however, fed from the pure blood of the vena cava and the aorta, is able to achieve complete development and is thus more perfect than the female seed.

It should be clear that this anatomical error by Galen provides the explanation, missing from the Aristotelian account, of why women are deficient in heat. Although Aristotle based all of woman's imperfections on this defect, he failed to provide an account of the mechanism that causes it. Galen's creative anatomy provides this mechanism. It is the impurity of the blood that generates the female seed that accounts for woman's inferior heat.

Galen, although rejecting the extreme Aristotelian position that women produce no seed, accepted and elaborated upon the belief in the primacy of the male role in generation. Galen developed one of the first biological explanations of the inferiority of the female creative principle. Since woman was conceived out of impure blood, she was colder than man. Because of her defect in heat, her organs of generation were not fully formed. Because her sexual organs were not fully developed, the seed produced by them would be imperfect. Religion had postulated a male being—Marduk, Yahweh, Zeus—as the ultimate creative force to account for the primacy of the male creative principle. Science pointed to the anatomy of the testes.

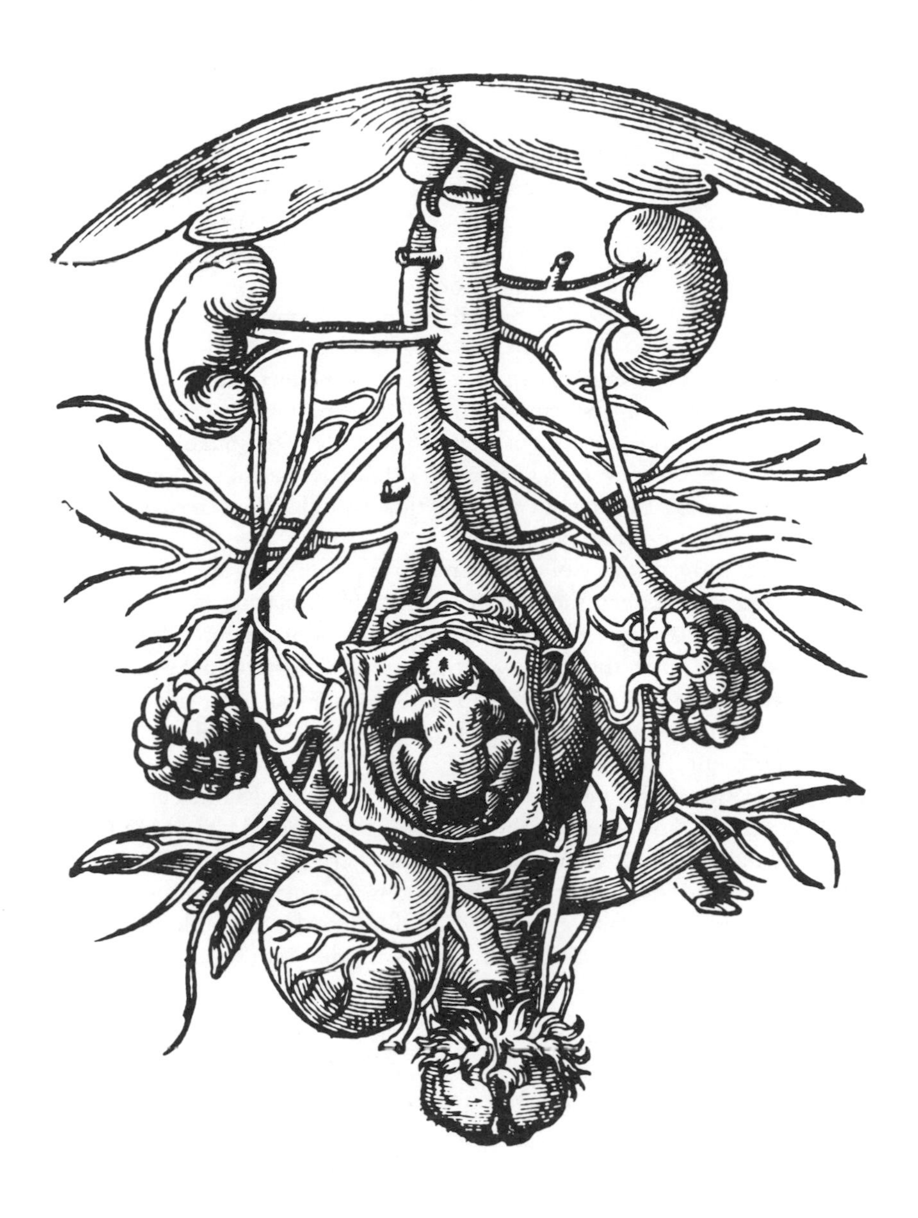

Illustration II. From Jacob Rueff (1580, 16). Notice the different points of origin of the vessels feeding the right and left ovaries.

THE PRIMACY OF MALE GENERATIVE POWERS
THE ANATOMICAL TRADITION

Aristotelian biology, corrected by the work of Galen, remained authoritative for many centuries. Aristotle's claim that man provided the primary causes of generation, and Galen's explanation of woman's imperfection and the related weakness of her creative powers were repeated by scientists well into the sixteenth century. Although the elegance of Galen's account might partially explain the popularity of this position, the fact that it is based on an anatomical error suggests an alternative explanation. Scientific theories reflect the values and attitudes of their authors. It is reasonable to believe that the assumption of woman's inferiority that has informed thought from the Classical period throughout modern time would influence scientific explanations as well as observations.

The belief in the primacy of male generative powers was an inherent part of the Western worldview. Viewing the creative force as male, both Christian and Jewish theorists had little trouble accepting and in turn reinforcing the Aristotelian view that the male was the primary creative force in human generation. Saint Gregory of Nyssa (331?-396), for example, adopting the theory of Aristotle, argued that the embryo was implanted by the male in the female. Saint Thomas Aquinas (1225?-1274) compared the generation of humans to that of the creation of the world. Just as God alone can produce a form in matter, "the active seminal power" of man's seed gives form to the "corporeal matter, which is supplied by the mother." Woman is capable only of the passive and less perfect role of providing the matter upon which the male semen will act (Aquinas 1947, I.98.2). Or consider the Midrashic image of generation which is identical to that of Aristotle. Man's sperm acts upon the woman's menstrual blood causing it to come together and giving it form.

> Job said "Has Thou not poured me out like milk and curdled me like cheese, . . ." (Job 10:10-12). A mother's womb is full of standing blood, which flows therefrom in menstruation. But, at God's will, a drop of whiteness enters and falls into its midst and behold a child is formed. This is likened unto a bowl of milk: when a drop of rennet falls into it, it congeals and stands; if not it continues as liquid. (Midrash Rabbah 14,19)

In the context of a system of beliefs in which the primacy of the male principle and the inferiority of women were such strongly imbedded beliefs, we can begin to understand why it took so many centuries for the anatomical error of Galen's explanation of the mechanism of woman's imperfection to be uncovered and accepted. I will demonstrate that the Western anatomical tradition from the thirteenth century to the sixteenth century perpetuated the views of Aristotle and Galen, making the same assumptions and the same errors.

In his *Anatomia Vivorum*, the anatomist Ricardi Anglici (1180-1225) relies on the authority of Aristotle. He insists that Aristotle was correct in thinking that male and female semen are intended for different purposes. He identifies female semen with menses and agrees with the Aristotelian position that female semen is so imperfectly developed that it can provide only the least perfect of all the causes of generation—the material cause.

> The purpose of the male sperm is to give form in the likeness of that from which the sperm comes; the purpose of the female sperm is to receive the likeness of that from which the sperm comes. From the male sperm, therefore, come spirit and creative power and form; from the female sperm come foundation, generation, and material. (Anglici 1927, 88)

Anglici employs a mold metaphor rather than the Aristotelian carpenter metaphor to describe the mechanism by which the sperm of the man gives form to the woman's sperm. "The male and female sperms mingle in the uterus, and the male sperm acts upon the female, for the male sperm naturally tends to impress the form of that from which it comes, and the female sperm tends to receive form" (Anglici 1927, 105). Once again the male seed is perceived as active, the female seed as passive.

Albertus Magnus (1206-1280), whose ideas were very influential and frequently copied, accepted the Galenic view that the female produces seed in addition to her menstrual fluids. Although accepting the existence of seed in women, Albertus allows it only a secondary role in generation. The seed of the man remains the active and formative agent, while the woman's seed merely prepares the menstrual fluids, the matter of generation, to receive the action of man's seed. The female seed has a role only in "preparing and enabling matter to receive the action from the operator, that is, man's sperm" (Albertus Magnus, xvi tr. 1 c. 16). The male sperm, endowed with the primary creative force, or what Albertus called "formative virtue," imparts to the female seed and the menstrual blood its form and end. Varying the carpenter metaphor of Aristotle to one similar to that of Anglici, Albertus offers the metaphor of imprinting to depict this process. The male sperm, upon entering the womb, "receives first the female sperm and then also the menstrual blood in which it stamps and imprints the creature's form and members" (Albertus Magnus xvi tr. 1 c. 10). Similar to Marduk forming the mountains and rivers out of the flesh of Tiamat or Yahweh giving form to the unformed earth, the male seed informs the blood and seed of the female.

The Galenic explanation of the cause of the weakness of female seed, and ultimately the cause of woman's inferior role in generation, was generally accepted in the fifteenth and sixteenth centuries. Alessandro Achillini (1463-1512), the Italian anatomist and philosopher, maintains that the sperm for the generation of the female passes through the left emulgent vein, which is "full of watery blood" that has yet to be cleansed in the kidney. Achillini claims to have performed, on two separate occasions, anatomical examinations in which he observed the left seminal vessel arising from the left emulgent vein. According to Achillini, such examinations also confirmed that the sperm for the generation of the male issue from the right vessel,

> because that vessel arises full of clean blood from the vena cava, that is, after the kidney sucks at the watery part, also because the branch of the artery is mingled with the right hand vessels and thus carries a more copious spirit. (Achillini [1520] 1975, 49)

Woman, generated from unclean blood that lacks a "copious spirit," cannot reach the perfection of being found in man.

Allesandro Benedetti (1450?-1512), a professor of medicine and anatomy at the University of Padua, developed a theory of generation that had an unusual blend of Aristotelian and Galenic premises. Similar to Aristotle he insists that woman's semen is "useless and vitiated." However, rather than identifying menses with woman's semen, Benedetti sides with Galen in claiming that women have semen in addition to menstrual fluids. He equates woman's semen with what he called "genital female whiteness." He also differs from Aristotle in holding that woman has a formative role in generation.

> The members of the foetus are formed from both the male and the female and the spiritual life is produced from both, but the principal members are constituted from the male seed and the other more ignoble ones from the female seed, as if from purer matter, just as the spiritual vigor is created." (Benedetti [1497] 1975, 99)

Thus, although Benedetti elevates woman's role in generation from that of Aristotle in allowing that she participates in the form of the offspring, her role remains clearly inferior to that of the male in that what she provides is less pure and forms only the secondary and "ignoble" members of the body.

The Spanish anatomist, Andres de Laguna (1499-1560), supported a position identical to that of Galen.

> The vessels which come to the left testis take their origin from the emulgent veins that are carried to the left kidney. Those which reach the right testis derive beyond doubt from the vena cava and the dorsal artery. For this reason the right testicle must be warmer than the left and the right side of the matrix warmer than the left side, especially since the more impure and serous blood always distills into the left testicle (de Laguna [1535] 1975, 278)

De Laguna argues that female seed is produced in the left testis. The impurity of the blood corrupts the seed resulting in the imperfection of the being so generated. This imperfection is manifested in the female organs of generation for, according to de Laguna, a woman produces little semen and what she does produce is much colder than that of man. Thus the female creative principle, though necessary, plays a secondary role in generation.

Dissection up until the sixteenth century was generally practiced with the object of illustrating the treatises of Galen or of the Moslem physician Avicenna (980-1037). However, as numerous discrepancies between these traditional texts and the conditions found in the cadavers were uncovered, there was a gradual realization among anatomists that the subject was in need of revision. The work of the anatomist Andreas Vesalius of Brussels (1514-1564) is widely recognized as central to this movement to question the tradition of authority and as founding modern anatomical practices.

Despite this atmosphere, anatomists persistently held to the view that the female seed was defective because of the impurity of the blood that fed it. Although careful attention to the actual structure of the veins and arteries of the testicles and ovaries would refute this view, anatomists continued to overlook this error. Even Vesalius did not perceive it. Notice in his depiction of the reproductive organs of women (see illustration III) that the vessel feeding the left ovary (e) originates in the renal artery (v) that carries uncleansed blood, while the right ovary is fed from the cleansed blood of the dorsal artery (d). It should not be concluded that this anatomical misconception was the result of ignorance of human female anatomy. Vesalius based his drawings of the female organs of reproduction on dissections of at least nine female bodies.

It is perhaps not surprising that even an anatomist as careful as Vesalius would perpetuate such an error. The scientific theory he had inherited demanded this "fact." The belief that female seed arose from the "serous, salty, and acrid" blood of the left testes was the only viable explanation of the perceived differences between women and men.

Niccolo Massa (1485-1569), the Italian doctor and anatomist, provides an interesting case study of the power of such biases. Massa appears to have recognized Galen's anatomical error. In his *Introductory Book of Anatomy* he claims that the vessels feeding the left ovary and testicle originate from the same source as those that feed the right.

> You will note, however, that often this left-hand vessel does not originate from the emulgent vein but from the trunk of the great chilis vein (vena cava) and also from the trunk of the aorta artery, just as the right-hand vessel does, as was said in the anatomy of the seminal vessels in males. (Massa [1559] 1975, 204)

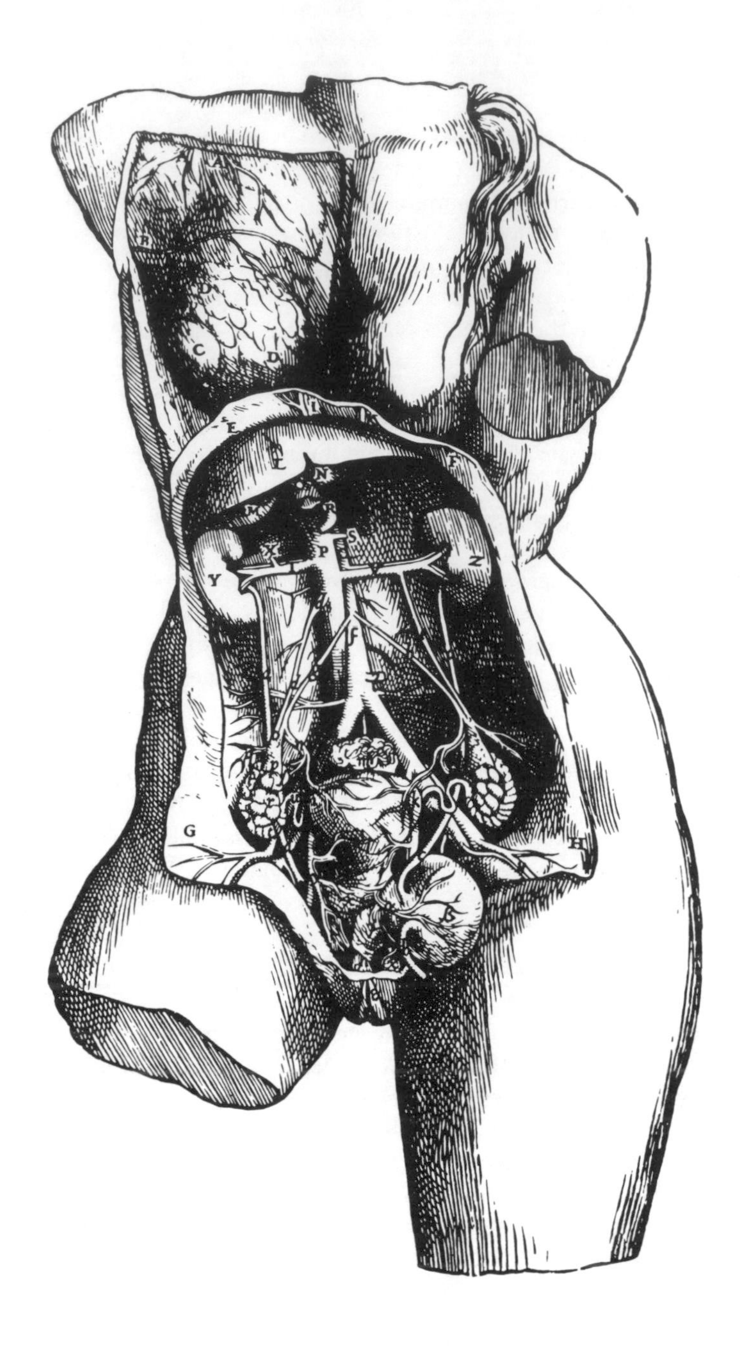

Illustration III. Vesalius. Plate 61 from *De Humani Corporis Fabrica.* Note the origins of the vessels feeding the right and left ovaries.

Notice his explicit rejection of the view that there is an anatomical difference between the seminal vessels of the right and left testicles and ovaries.

Despite this, Massa continues to insist that female seed is weaker than male seed. Amazingly, his explanation of the imperfection of female seed is identical to that of Galen! "I also know that to the left testicle of the woman as well as of the man a watery blood is sent from the emulgents. From this a sperm is created which is not very strong" (Massa [1559] 1975, 207). Although able to see the error of Galen's anatomy, Massa is unable to perceive its implications. He continues to support Galen's theoretical explanation of woman's inferiority, while having undercut the physiological basis for it.

The image of woman as an incomplete or inferior man contributed to the persistence of the Galenic explanation of the cause of this difference. Despite the "corrections" of anatomists like Massa, it would take over a century for Galen's error to be generally recognized. The physician Hermann Boerhaave (1668-1738), for example, at the end of the seventeenth century, would admit that he too was initially "seduced" by authority into believing that the vessels feeding the left testicle and ovary originated from a different source than those of the right. Although acknowledging the mistake, he proceeds to "explain" it with yet another piece of creative anatomy, claiming that the veins and arteries of the left ovary and testicle are wrapped in a "capsule" or "tunic" that makes their point of origin more difficult to see than with the right ovary and testicle (Boerhaave 1757, 42n). The strength of the bias of woman's inferiority made such a convenient explanation difficult to give up, despite its inherent error.

THE SEEDS OF CREATION: PREFORMATION DOCTRINE

For almost 2000 years science consistently viewed woman's contribution to human generation as less potent than man's. In the seventeenth century a new theory took hold that had significant effects on the science of embryology. From Aristotle to Harvey, embryologists had viewed the embryo as produced through gradual development from unorganized matter, what we now call an epigenetic theory of development. Form was seen as evolving from the actions of the male semen upon the blood of the uterus or from the mixture of the semen of the two parents. The idea of an evolution of complexity from unstructured material lost favor toward the end of the seventeenth century as a result of the general scientific commitment to a mechanistic worldview and the insufficiency of mechanical explanations of the gradual development of living organisms. The epigenetic view was rejected and replaced by the thesis that development was the result of the growth or unfolding of pre-existing structures—preformation doctrine (See Gasking 1967). The theory of preformation was initially formulated in the works of Malebranche, Perrault, and Swammerdam. In 1669, the entomologist Jan

Swammerdam (1637-1680) claimed that "there is never generation in nature, but only a lengthening or increase in parts" (Swammerdam 1672, 21). Fetal development was likened to the enlargement of the little leaves in a bud. The structures of the entity were all there from the beginning. Growth increased only the size, never the complexity, of the pre-existing parts. For approximately a century, beginning in the last quarter of the seventeenth century, preformation doctrine supplanted epigenetic theory.

According to early versions of preformation theory, the spot in the center of a chicken's egg contained, preformed, all of the essential structures of a chick. Its parts were, to be sure, far too small to be seen. The anatomist Marcello Malpighi (1628-1694), utilizing microscopic magnification, supported this theory in his study of chick embryos. He observed and prepared plates showing the early states of the heart, liver, vertebrae, and other structures. He observed that a small, but fully formed heart was visible in the embryo well in advance of the time it began to beat, thus lending credence to the view that the chick began fully formed, although the size of each part had to increase prior to becoming visible.

If it is not possible for an entity to increase in complexity, then embryos must exist fully formed within the parent. In fact, carried to its logical conclusion, the theory of preformation would require that each seed itself must contain within it the seeds of subsequent generations. This view came to be known as encasement. It was held that Adam and Eve contained all humanity in their loins, each generation encased in the seed of the previous generation.

An important consequence of preformation theory should, by now, be obvious. The seed of subsequent generations could reside only in a single parent. Generation could no longer be the result of a mixture of semen. The only question was whether the miniature being was in the female egg (ovism) or in the male sperm (animalculism). Preformationists initially saw the female as the source of this seed and argued that the male parent provided only a stimulus that caused the outermost seed to begin its growth. However, by 1680 a version of preformation was offered that viewed the male parent as the source of seed.

Chicken eggs were a popular subject of study for embryologists for obvious reasons. The eggs were large enough to allow for observation, as well as being readily and cheaply available. Many embryologists, from Aristotle to Malpighi, had grounded much of their empirical knowledge of embryology upon the study of chick embryos. Given this, it is not surprising that preformationists first turned to the female as the source of encased seed. Spermatozoa had yet to be observed in any animal, while scientists had numerous examples of female animals that contained visible eggs.

This situation changed when Louis Dominicus Hamm, the assistant of Anthony van Leeuwenhoek (1632-1723), while looking at human seminal emissions under a microscope, saw "spermatic animalcules." In a series of correspondences with the Royal Society beginning in 1677, Leeuwenhoek claimed to have confirmed the discovery of Hamm and concluded that these animalcules were the source of embryos. To counter the arguments of ovists, he studied female ovaries and concluded that the mammalian ovaries were useless ornaments. Echoing centuries of tradition, Leeuwenhoek insisted that the nourishment of the masculine seed was the sole function of the female. To account for the generation of the two sexes, Leeuwenhoek claimed to have observed two kind of spermatozoa, one that would give rise to males, the other to females (Leeuwenhoek 1685, 1120-34).

Nicolaus Hartsoeker (1656-1725), in 1678, was the first scientist to illustrate the appearance of a fetus contained within a spermatozoon. (See illustration IV.) Hartsoeker did not claim to have seen such a being, but insisted that if we could see through the skin which hides it, we might see it as he represented it in the illustration.

Those theorists who supported animalculism, the animal or all the essential structures of the animal preformed in the spermatozoon, over ovism, the animal preformed in the egg, saw the egg as simply a temporary abiding place for the spermatic animalcule that provided it with food, shelter, and warmth. On this perspective, an organism's entire structure resides in the sperm, but the female egg is still necessary as a nutritive mass. Animalculists argued that only the sperm that is able to find the egg and pierce it will begin the growth process.

Although ovists were initially without rivals, the theory quickly went out of favor once the male analogue to the egg was observed. Within two decades, animalculism was the favored view, and continued to be held for almost a century. By 1730 ovism was almost uniformly rejected, although it was revived for a short period from 1760-1780 by Haller, Bonnet, and Spallanzani just prior to the return to epigenetic theories. By the end of the eighteenth century, preformation doctrine was replaced by epigenetic theory (see Roe 1981).

For our purposes, what is fascinating about the history of preformation doctrine is the general popularity of animalculism over ovism. Clearly, some of the popularity of animalculism was the centuries long bias of male primacy in generation. On the ovist view, the male merely provides the stimulus that initiates the unfolding process, but it was the female that contained the formed being—a theory that lent itself to viewing the female as contributing more to generation than the male. There were, to be sure, problems with both theories. Ovists were faced with the difficult question of how a female egg could travel from the ovary to a fallopian tube, as well as how it could pass through the initial narrows into the uterus. The animalculists were faced with a more serious problem than ovists, namely the huge numbers of spermatozoa in each emission. Believing that each spermatozoon contained a miniscule individual, itself containing generations of such individuals, meant that

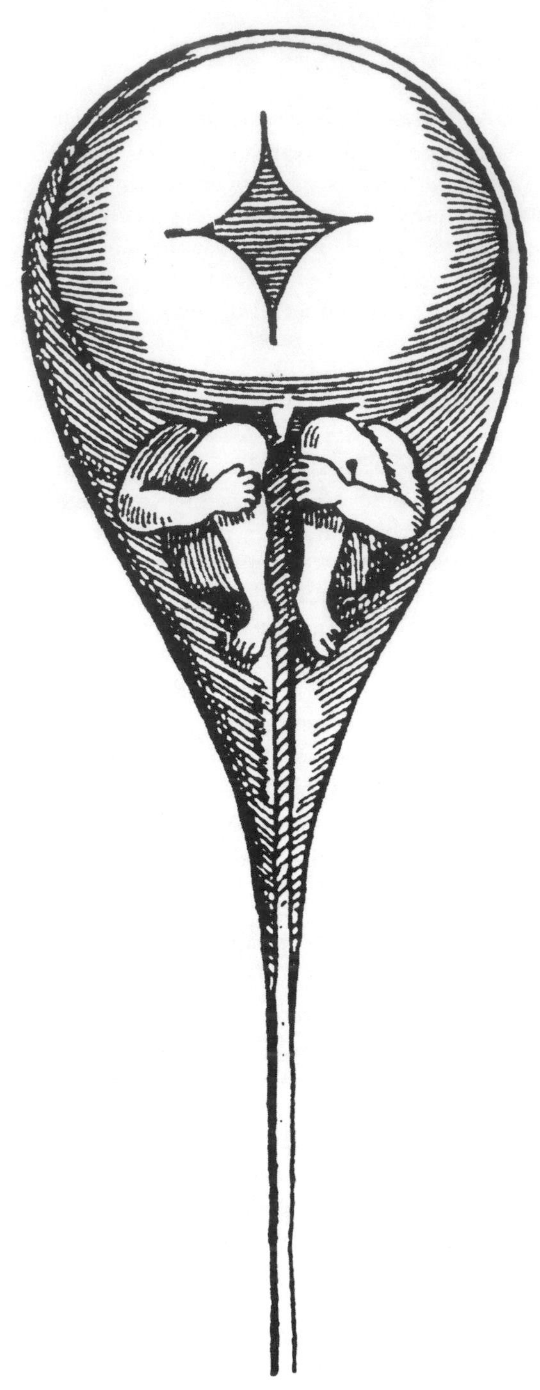

Illustration IV. Drawn by Nicolaus Hartsoeker in his *Essai de diotropique* (1694). The figure was to represent what Hartsoeker believed would be seen if a sperm could be viewed without its protective covering.

hundreds of such individuals had to be sacrificed in order for one of them, at best, to achieve the conditions necessary for growth. This posed a problem since it was difficult to understand why a godly designed universe would contain such waste. The popularity of animalculism in the face of this suggests the strength of the bias of male primacy in creation.

Although the animalculist theory was in sharp contrast to Aristotelian embryological theories in rejecting epigenesis for preformation, animalculist premises sound surprisingly similar to those of Aristotle. The animalculist George Garden argued that the egg, although essential to development, provided none of the form of the fetus. The ovum supplied only the nutriment necessary for the unfolding of the embryo already fully formed in the sperm (Garden 1691, 474-83). "The Animalcules of the male Semen contain," according to Boerhaave, "the future Rudiments of the whole body." The female supplies the "nidus" or egg that will offer nourishment and warmth enabling the animalcule to unfold itself "so as to display the latent Parts, of which [it is] composed" (Boerhaave 1757, 138).

Martin Frobenius Ledermuller (1719-1769) claimed that "the small seminal animalcule develops in the mother just as the seed does in the field" (Cole 1930, 108). J. Cooke in 1762, compared animalcules to seeds of plants that are nourished by the egg as is the seed by the "juices of the earth." In an image echoing Aeschylus' Apollo, Cooke insists that animalculism is

> confirmed by the natural Instinct of Mankind (for the Children are not denominated from the Mother, which if the Stamina had preexisted in her Eggs, and not in the Male Sperm, they in Strictness ought to have been); but they are, as in Justness they should be denominated from the Father, as proceeding first of all out of his Body, before by Birth they came out of the Body of their Mother. (Cooke 1762, 63-4)

Although one of the later animalculists, Cooke continued to hold to Hartsoeker's metaphor of a fetus covered by skin. The spermatozoa carried by a man "are no less than his own little Pupilla, Images, or Pictures in Parvo, wrapt up very securely, Insect like, in a fine exterior Bag, Covering, or Wrapper" (Cooke 1762, 15-16). Woman continues to play a role in generation, but one simply of feeder and caretaker of the growing animalcule.

Even Erasmus Darwin (1731-1802), as late as 1794, insists that the male provides the form or rudiment of the embryo, while the female provides only the oxygen, food, and nindus. To support this position, Darwin reminds us of the larger size of the male and insists that given his larger size he "should contribute as much or more towards the reproduction of the species." Since the female provides food and oxygen, she could not, on this logic, also provide part of the embryon or her contribution would be greater than that of the male. From this Darwin concludes "that the embryon is produced by the male, and the proper food and nidus by the female" (Darwin 1794, 484).

In an interesting deviation from the Aristotelian carpenter metaphor, animalculists employed a mold metaphor to account for the fact that some children resemble their mothers. Jean Astruc (1686-1766), for example, argued that

> the worm which is the germ of the foetus, comes from the father only; the mother has no part in it . . . in the little nich of the ova of women, there is a concave impression; that resembles every woman; and is placed in every nich, on the same side, as where the end of the cord is: which must be the ground of the resemblance of children to their mothers. (Astruc 1762, 47-48)

As in the creation of Pandora, the female creative principle was given a role only in the external appearance of the created being. All else was attributed to the male.

CONCLUSION

The history of embryology from Aristotle to the preformationists provides an excellent illustration of the ways in which what Evelyn Fox Keller calls the gender/science system informs the process of scientific investigation and theory formation. The accepted belief in woman's inferiority can be shown to have affected, in various scientists, the process of observation, the interpretation of data, and the justification and defense of theory.

Aristotle set the basic orientation for the next 2000 years of embryological thought. His arguments, supplemented by the work of Galen, were retained even in the face of growing evidence of their untenability. There were basically two lines of development: 1) Aristotle and the denial of female seed, and 2) Galen and the role of inferior female seed. The anatomical tradition followed Galen, while the animalculists, although offering a very different view of the mechanism of generation, continued to employ images of the role of the female similar to those of Aristotle and Galen. Throughout, the basic belief in the primacy of the male remained virtually unchallenged.

The history of Western views of human generation provides more than a chronicle of the legacy of Aristotle and Galen. The tenacious defense of the belief in the primacy of the male generative powers in the face of growing evidence against it, reveals the deep-rooted nature of this conviction and the emotional valence attached to it. Aristotle and Galen were key figures, but part of their prominence comes through their articulation of a systematic explanation that preserved and defended this conviction. Their influence cannot be explained solely in terms of their role as authoritative figures. In addition, it is necessary to recognize that the survival and renewal of their views was, to a significant degree, due to their legitimation of the deeply held belief in woman's inferiority.

Lest one be tempted to think that such biases vanished or were at least minimized with the rise of modern science and the refinement of the experimental method, one has only to turn the page to the essay written by the Biology and Gender Study Group (this volume). Through their excellent analysis we see that during the ninteenth and twentieth century the gender bias of biology continued with vigor.

The gender/science system is woven tightly into the fabric of science. To unravel the complexities of the pattern of bias against women and reweave our theories and practice of science will require a similar transformation of our worldview and of the social practices and institutions that are justified by the gender/science system and in turn reinforce it.

References

Aeschylus. 1975. *The oresteia.* Trans. Robert Fagles. New York: Viking Press.

Albertus Magnus. n.d. *De animalibus.* Trans. J. Quincy. Albert von Bollstadt.

Achillini, Allesandro. [1520] 1975. *Anatomical notes.* Trans. L.R. Lind. In *Studies in pre-Vesalian anatomy: Biography, translations, documents*, ed. L. R. Lind. Philadelphia: The American Philosophical Society.

Anglici, Ricardi. 1927. *Anatomia vivorum.* Trans. George Corner. In *Anatomical texts of the earlier middle ages*, ed. George Corner. Washington: Carnegie Institution.

Aquinas, Thomas. 1947. *Summa theologica.* Trans. Fathers of the English Dominican Provinces. New York: Benziger Brothers.

Aristotle. 1984. *The generation of animals.* Trans. A. Platt. In *The complete works of Aristotle*, Vol. 8, ed. J. Barnes. Princeton: Princeton University Press.

————1984. *The history of animals.* Trans. A.W. Thompson. In *The complete works of Aristotle*, Vol. I, ed. J. Barnes. Princeton: Princeton University Press.

Astruc, J. 1762. *A treatise on the diseases of women*, Vol. III. London: J. Nourse, Bookseller.

Benedetti, Allesandro. [1497] 1975. *History of the human body.*Trans. L.R. Lind. In *Studies in pre-Vesalian anatomy: Biography, translations, documents*, ed. L. R. Lind. Philadelphia; The American Philosophical Society.

Boerhaave, Herman. 1757. *Dr. Boerhaave's academical lectures on the theory of physic.* London.

Cole, F.J. 1930. *Early theories of sexual generation.* Oxford: Clarendon Press.

Cooke, J. 1762. *The new theory of generation.* London: J. Buckland.

Darwin, Erasmus. 1794. *Zoonomia: Or the laws of organic life*. New York: AMS Press.

Dobell, Clifford. 1960. *Anthony von Leeuwenhoek and his little animals*. New York: Dover

Galen. 1968. *On the usefulness of the parts of the body*. Trans. M.T. May. Ithaca: Cornell University Press.

Garden, George. 1691. A discourse concerning the modern theory of generation. *Philosophical Transactions of the Royal Society*. Vol. XVII.

Gasking, Elizabeth. 1967. *Investigations into generation: 1651-1828*. Baltimore: Johns Hopkins University Press.

Hartsoeker, Nicolaus. 1694. *Essai de diotropique*.

Hippocrates. 1943. *Regimen*. Trans. W.H.S. Jones. Cambridge: Harvard University Press.

Horowitz, Maryanne Cline. 1976. Aristotle and women. *Journal of the History of Biology* 9 (2): 183-213.

de Laguna, Andres. [1535] 1975. *Anatomical procedure, or a survey of the dissections of the human body*. Trans. L.R. Lind. In *Studies in pre-Vesalian anatomy: Biography, translations, documents*, ed. L.R. Lind. Philadelphia: The American Philosophical Society.

Lange, L. 1983. Woman is not a rational animal: On Aristotle's biology of reproduction. In *Discovering reality*, ed. S. Harding and M. Hintikka. Dordrecht: D. Reidel.

Leeuwenhoek, Anthony van. 1685. *Philosophical transactions of the Royal Society* XXV: 1120-34.

Malebranche, Nicolas. 1965. 252 entretiens sur la metaphysique et sur la religion. In *Oeuvres completes* Vol. 12. Paris: Librairie Philosophique J. Vrin.

Massa, Niccolo. [1559] 1975. *Introductory book of anatomy*. Trans. L.R. Lind. In *Studies in pre-Vesalian anatomy: Biography, translations, documents*, ed. L.R. Lind. Philadelphia: The American Philosophical Society.

Meyer, A.W. 1956. *Human generation: Conclusions of Burdach, Dollinger and von Baer*. Stanford: Stanford University Press.

Midrash Rabbah. 1921. Vilna: Romm. Cited in *Marital relations, birth control, and abortion in Jewish law* by David M. Feldman, 133. New York: Schocken Books.

Needham, Joseph. 1934. *A history of embryology*. Cambridge: Cambridge University Press.

Paré, Ambroise. 1968. *The collected works*. Trans. T. Johnson. New York: Milford House.

Preus, Anthony. 1975. *Science and philosophy in Aristotle's biological works*. New York.

———. 1977. Galen's criticism of Aristotle's conception theory. *Journal of the History of Biology* 10 (1):65-85

Roe, Shirley. 1981. *Matter, life, and generation: Eighteenth-century embryology and the Haller-Wolff debate*. Cambridge: Cambridge University Press.

Rueff, Jacob. 1580. *De conceptu et generatione hominis*. Frankfort.

Ruestow, Edward. 1983. Images and ideas: Leeuwenhoek's perception of the spermatozoa. *Journal of the History of Biology* 16 (2):185-224

Swammerdam, Jan. 1672. *Miraculum naturae*.

Vesalius, Andreas. 1949. *The epitome of Andreas Vesalius*. Trans. L.R. Lind. Cambridge: MIT Press.

———. 1950. *The illustrations from the works of Andreas Vesalius of Brussels*. Ed. and trans. J.B. de C.M. Saunders and Charles D. O'Malley. Cleveland: World Publishing Co.

The Importance of Feminist Critique for Contemporary Cell Biology

THE BIOLOGY AND GENDER STUDY GROUP
Athena Beldecos, Sarah Bailey, Scott Gilbert, Karen Hicks
Lori Kenschaft, Nancy Niemczyk, Rebecca Rosenberg
Stephanie Schaertel, and Andrew Wedel

Biology is seen not merely as a privileged oppressor of women but as a co-victim of masculinist social assumptions. We see feminist critique as one of the normative controls that any scientist must perform whenever analyzing data, and we seek to demonstrate what has happened when this control has not been utilized. Narratives of fertilization and sex determination traditionally have been modeled on the cultural patterns of male/female interaction, leading to gender associations being placed on cells and their components. We also find that when gender biases are controlled, new perceptions of these intracellular and extracellular relationships emerge.

Nancy Tuana (this volume) has traced the seed-and-soil analogy from cosmological myths through Aristotle into the biology of the 1700s. Modeling his embryology after his social ideal, Aristotle promulgated the notions of male activity versus female passivity, the female as incomplete male, and the male as the real parent of the offspring. The female merely provided passive matter to be molded by the male sperm. While there were competing views of embryology during Aristotle's time, Aristotle's principles got the support of St. Thomas and were given the sanction of both religion and scientific philosophy (Horowitz 1976, 183). In this essay, we will attempt to show that this myth is still found in the core of modern biology and that various "revisionist" theories have been proposed within the past five years to offset this myth.

We have come to look at feminist critique as we would any other experimental control. Whenever one performs an experiment, one sets up all the controls one can think of in order to make as certain as possible that the result obtained does not come from any other source. One asks oneself what assumptions one is making. Have I assumed the temperature to be constant? Have I assumed that the pH doesn't change over the time of the reaction? Feminist critique asks if there may be some assumptions that we haven't checked concerning gender bias. In this way feminist critique should

Hypatia vol. 3, no. 1 (Spring 1988).

be part of normative science. Like any control, it seeks to provide critical rigor, and to ignore this critique is to ignore a possible source of error.

The following essay is not an attempt to redress past injustices which biology has inflicted upon women. This task has been done by several excellent volumes that have recently been published (Sayers 1982; Bleier 1984, 1986; Fausto-Sterling 1985). Rather, this paper focuses on what feminist critique can do to strengthen biology. What emerges is that gender biases do inform several areas of modern biology and that these biases have been detrimental to the discipline. In other words, whereas most feminist studies of biology portray it—with some justice—as a privileged oppressor, biology has also been a victim of the cultural norms. These masculinist assumptions have impoverished biology by causing us to focus on certain problems to the exclusion of others, and they have led us to make particular interpretations when equally valid alternatives were available.

Sperm Goes A'Courtin'

If Aristotle modeled fertilization and sex determination on the social principles of his time, he had plenty of company among more contemporary biologists. The first major physiological model of sex determination was proposed in 1890 when Sir Patrick Geddes and J. Arthur Thomson published *The Evolution of Sex*, one of the first popular treatises on sexual physiology. By then, it had been established that fertilization was the result of the union of sperm and egg. But still unanswered was the mechanism by which this event constructed the embryo. One of the central problems addressed by this highly praised volume was how sex was determined. Their theory was that there were two types of metabolism: *anabolism*, the storing up of energy, and *katabolism*, the utilization of stored energy. The determination of sexual characteristics depended on which mode of metabolism prevailed. "In the determination of sex, influences favoring katabolism tend to result in the production of males, as those favoring anabolism similarly increase the production of females" (Geddes and Thomson 1890, 45, 267). This conclusion was confirmed by looking at the katabolic behavior of adult males (shorter life span, greater activity and smaller size) compared to the energy-conserving habits of females who they described as "larger, more passive, vegetative, and conservative."[1] In a later revision (1914, 205-206) they would say, "We may speak of women's constitution and temper as more conservative, of man's more unstable. . . . We regard the woman as being more anabolic, man as relatively katabolic; and whether this biological hypothesis be a good one or not, it certainly does no social harm."

This microcosm/macrocosm relationship between female animals and their nutritive, passive eggs and between male animals and their mobile, vigorous sperm was not accidental. Geddes and Thomson viewed the sperm and egg as representing two divergent forms of metabolism established by protozoan organisms, and "what was decided among the prehistoric protozoa cannot be annulled by Act of Parliament." Furthermore, as in Aristotle, the difference between the two is nutrition. The motivating force impelling the sperm towards the egg was hunger. The yolk-laden egg was seen as being pursued by hungry sperm seeking their nourishment. The Aristotelian notion of activity and passivity is again linked with the role of female as nutrient provider. It is also linked with that most masculine of British rituals, the hunt.[2]

It is usually assumed that the discovery of the X and Y sex chromosomes put an end to these environmental theories of sex determination. This is today's interpretation and not that of their discoverer. What the genetics texts do not tell us is that C.E. McClung placed his observations of sex chromosomes directly in the context of Geddes and Thomson's environmental model. Using a courtship analogy wherein the many spermatic suitors courted the egg in its ovarian parlour, McClung (1901, 224) stated that the egg "is able to attract that form of spermatozoon which will produce an individual of the sex most desirable to the welfare of the species." He then goes on to provide an explicit gender-laden correlation of the germ cells mirroring the behavior of the sexual animals that produced them:

> The ovum determines which sort of sperm shall be allowed entrance into the egg substance. In this we see the extension, to its ultimate limit, of the well-known role of selection on the part of the female organism. The ovum is thus placed in a delicate adjustment with regard to the surrounding conditions and reacts in a way to best subserve the interests of the species. To it come two forms of spermatozoa from which selection is made in response to environmental necessities. Adverse conditions demand a preponderance of males, unusually favorable conditions induce an excess of females, while normal environments apportion an approximately equal representation of the sexes. (McClung 1902,76)

McClung concluded this paper by quoting that Geddes and Thomson's theory of anabolism and katabolism provided the best explanation as to whether the germ cells would eventually grow into "passive yolk-laden ova or into minute mobile spermatozoa."

THE SPERM SAGA

Courtship is only one of the narrative structures used to describe fertilization. Indeed, "sperm tales" make a fascinating subgenre of science fiction. One of the major classes of sperm stories portrays the sperm as a heroic victor. In these narratives, the egg doesn't choose a suitor. Rather, the egg is the passive prize awarded to the victor. This epic of the heroic sperm struggling against the hostile uterus is the account of fertilization usually seen in contemporary introductory biology texts. The following is from one of this decade's best introductory textbooks.

> Immediately, the question of the fertile life of the sperm in the reproductive tract becomes apparent. We have said that one ejaculation releases about 100 million sperm into the vagina. Conditions in the vagina are very inhospitable to sperm, and vast numbers are killed before they have a chance to pass into the cervix. Millions of others die or become infertile in the uterus or oviducts, and millions more go up the wrong oviduct or never find their way into an oviduct at all. The journey to the upper portion of the oviducts is an extremely long and hazardous one for objects so tiny. . . . Only one of the millions of sperm cells released into the vagina actually penetrates the egg cell and fertilizes it. As soon as that one cell has fertilized the egg, the [egg] cell membrane becomes impenetrable to other sperm cells, which soon die. (Keeton 1976,394)

We might end the saga by announcing, "I alone am saved." These sperm stories are variants of the heroic quest myths such as the Odyssey or the Aeneid. Like Aeneas, the spermatic hero survives challenges in his journey to a new land, defeats his rivals, marries the princess and starts a new society. The sperm tale is a myth of our origin. The founder of our body is the noble survivor of an immense struggle who deserved the egg as his reward. It is a thrilling and self-congratulatory story.

The details of these fertilization narratives fit perfectly into Campbell's archetype of such myths. Campbell (1956,387), however, believes that "there is no hiding place for the gods from the searching telescope or microscope." In this he has been wrong. The myth lies embedded within microscopic science.[3]

The next passage comes from a book to be given expectant mothers. It, too, starts with the heroic sperm model but then ventures off into more disturbing images.

> Spermatozoa swim with a quick vibratory motion. . . . In ascending the uterus and Fallopian tube they must swim against the same current that waft the ovum downward. . . . Although a million spermatozoa die in the vagina as a result of the acid secretions there, myriads survive, penetrate the neck of the uterus and swarm up through the uterine cavity and into the Fallopian tube. *There they lie in wait for the ovum.* As soon as the ovum comes near the *army of spermatozoa*, the latter, as

> if they *were tiny bits of steel drawn by a powerful magnet, fly at the ovum*. One *penetrates*, but only one. . . . *As soon as the one enters, the door is shut on other suitors*. Now, as if *electrified*, all the particles of the ovum (now fused with the sperm) exhibit vigorous agitation. (Russell 1977, 24, emphasis added)

In one image we see the fertilization as a kind of martial gang-rape, the members of the masculine army lying in wait for the passive egg. In another image, the egg is a whore, attracting the soldiers like a magnet, the classical seduction image and rationale for rape. The egg obviously wanted it. Yet, once *penetrated*, the egg becomes the virtuous lady, closing its door to the other *suitors*. Only then is the egg, because it has fused with a sperm, rescued from dormancy and becomes active. The fertilizing sperm is a hero who survives while others perish, a soldier, a shard of steel, a successful suitor, and the cause of movement in the egg. The ovum is a passive victim, a whore and finally, a proper lady whose fulfillment is attained.

The accounts in such textbooks must seem pretty convincing to an outsider. The following is from a paper on the history of conception theories, published—by a philosopher—in 1984.

> Aristotle's intuitions about the male as trigger which begins an epigenetic process is a foreshadowing of modern biological theory in which the sperm is the active agent that must move and penetrate the ovum. The egg passively awaits the sperm, which only contributes a nucleus, whereas the egg contributes all the cytoplasmic structures (along with its nucleus) to the zygote. In other words, the egg contributes the material and the form, and the sperm contributes the activating agent and the form. . . . Thus even modern biology recognizes the specialized and differentiated roles of male and female in an account of conception. Aristotle's move in such a direction was indeed farsighted. (Boylan 1984, 110)

ENERGETIC EGGS AND ACTIVE ANLAGEN

Until very recently, textbook accounts have emphasized (even idealized) the passivity of the egg. The notion of the male semen "awakening the slumbering egg" is seen as early as 1795 (Reil 1795, 79), and this idea, according to historian Tim Lenoir (1982, 37) "was to have an illustrious future." Since 1980, however, there has been a new account of sperm-egg interactions. This revisionism has been spurred on by new data (and new interpretations of old data) which has forced a re-examination of the accepted scenario. The egg appears to be less a "silent partner" and more an energetic participant in fertilization. Two of the major investigators forcing this re-evaluation are

Gerald and Heide Schatten. Using scanning electron microscopy, they discovered that when the sperm contacts the egg, it does not burrow through.[4] Rather, the egg directs the growth of microvilli—small finger-like projections of the cell surface—to clasp the sperm and slowly draw it into the cell. The mound of microvilli extending to the sperm had been known since 1895 when E. B. Wilson published the first photographs of sea urchin fertilization. But this structure has been largely ignored until the recent studies, and its role is still controversial.

In 1983, the Schattens wrote a review article for laypeople on fertilization. Entitled "The Energetic Egg," it consciously sought to change the metaphors by which fertilization is thought about and taught.

> In the past years, investigations of the curious cone that Wilson recorded have led to a new view of the roles that sperm and egg play in their dramatic meeting. The classic account, current for centuries, has emphasized the sperm's performance and relegated to the egg the supporting role of Sleeping Beauty—a dormant bride awaiting her mate's magic kiss, which instills the spirit that brings her to life. The egg is central to this drama, to be sure, but it is as passive a character as the Grimm brothers' princess. Now, it is becoming clear that the egg is not merely a large yolk-filled sphere into which the sperm burrows to endow new life. Rather, recent research suggest the almost heretical view that sperm and egg are mutually active partners. (Schatten and Schatten 1983,29)

Other studies are showing this mutual activity in other ways. In mammals, the female reproductive tract is being seen as more than a passive or even hostile conduit through which sperm are tested before they can reach the egg. Freshly ejaculated mammalian sperm are not normally able to fertilize the eggs in many species. They have to become *capacitated*. This capacitation appears to be mediated through secretions of the female genital tract. Furthermore, upon reaching the egg, mammalian sperm release enzymes which digest some of the extracellular vestments which surround the egg. These released enzymes, however, are not active. They become activated by interacting with another secretion of the female reproductive tract. Thus, neither the egg nor the female reproductive tract is a passive element in fertilization. The sperm and the egg are both active agents and passive substrates. "Ever since the invention of the light microscope, researchers have marveled at the energy and endurance of the sperm in its journey to the egg. Now, with the aid of the electron microscope, we can wonder equally at the speed and enterprise of the egg, as it clasps the sperm and guides its nucleus to the center" (Schatten and Schatten 1983, 34).

As we have seen above, the determination of maleness and femaleness has also been inscribed by concepts of active masculinity and passive femaleness. (This means that *sex*, not just *gender*, can be socially constructed!) Indeed, until 1986, all modern biological theories of mammalian sex determination have assumed that the female condition is developed passively, while the male condition is actively produced from the otherwise female state (for review, see Gilbert 1985, 643). This has been based largely on Jost's experiments where rabbits developed the female body condition when their gonadal rudiments were removed before they had differentiated into testes or ovaries. But these experiments actually dealt with the generation of secondary sexual characteristics and not the primary sex determination event—the differentiation of the sexually indifferent gonadal primordia into ovaries or testes.

During the past four years, these theories of primary sex differentiation (notably the H-Y antigen model wherein male cells synthesized a factor absent in female cells which caused the gonadal primordia to become testes) have been criticized by several scientists, and a new hypothesis has been proposed by Eva Eicher and Linda Washburn of the Jackson Laboratory. This new model is based on extensive genetic evidence and incorporates data that could not be explained by the previous accounts of sex determination. In their introductory statement, Eicher and Washburn point out the active and passive contexts that have been ascribed to the development of the primary sexual organs. They put forth their hypotheses as a controlled corrective for traditional views.

> Some investigators have over-emphasized the hypothesis that the Y chromosome is involved in testis determination by presenting the induction of testicular tissue as an active (gene directed, dominant) event while presenting the induction of ovarian tissue as a passive (automatic) event. Certainly, the induction of ovarian tissue is as much an active, genetically directed developmental process as is the induction of testicular tissue or, for that matter, the induction of any cellular differentiation process. Almost nothing has been written about genes involved in the induction of ovarian tissue from the undifferentiated gonad. The genetics of testis determination is easier to study because human individuals with a Y chromosome and no testicular tissue or with no Y chromosome and testicular tissue, are relatively easy to identify. Nevertheless, speculation on the kind of gonadal tissue that would develop in an XX individual if ovarian tissue induction fails could provide criteria for identifying affected individuals and thus lead to the discovery of ovarian determination genes. (Eicher and Washburn 1986, 328)

Again, we see that alternative versions of long-held scientific "truths" can be generated. A feminist critique of cellular and molecular biology does not necessarily mean a more intuitivistic approach. Rather, it involves being open to different interpretations of one's data and having the ability to ask questions that would not have occurred within the traditional context. The studies of Eicher and Washburn on sex determination and those of the Schattens on fertilization can be viewed as feminist-influenced critiques of cell and molecular biology. They have controlled for gender biases rather than let the ancient myth run uncontrolled through their interpretations. Yet the techniques used in their analyses are not different than those of other scientists working in their respective fields, and the approaches used in these studies are no "softer" than those used by researchers working within the traditional paradigms.[5]

A Nuclear Family: The Sexualization of the Cell

The sperm and egg are *gametes*; that is marriage partners. As we have seen, their interactions have been modeled on various courtship behaviors. This extrapolates, however, into a husband-wife arrangement in the zygote cell. It is again not surprising, then, to find this relationship reflected in the relationship between nucleus and cytoplasm. The sperm, after all, is viewed as a motile nucleus while the cytoplasm of the zygote and its descendants is derived entirely from the ovum (Morgan 1926, 45). One might argue that the ovum provides a nuclear component equal to that of the sperm, but this is usually overlooked (note the parentheses in the above quotation from Boylan). Even today among biologists, the term "maternal inheritance" is identical with "cytoplasmic inheritance." The nucleus came to be seen as the masculine ruler of the cell, the stable yet dynamic inheritance from former generations, the unmoved mover, the mind of the cell. The cytoplasm became the feminine body of the cell, the fluid, changeable, changing partner of the marriage.

This marriage trope was extremely prevalent during the 1930's when there were at least four competing views of the relationship between the cytoplasm and the nucleus (Gilbert, in press). What one finds is that the relationship of husband to wife becomes that of nucleus to cytoplasm. In Germany, one of the dominant theories modeled the cell after an autocratic Prussian family. The nucleus contained all the executive functions and the cytoplasm did whatever the nucleus commanded. Indeed, the cytoplasm existed only to be physically acted upon by the nuclear genes. As Harwood (1984, 3) has pointed out, defenders of this *Kernmonopol* wrote of the supremacy ("*Uberlegenheit*") of the genes and the dominating role of the nucleus ("*die dominierende Rolle des Kernes*"). The leading American geneticist, T.H. Morgan, modeled the cell after a more American family. First, the nucleus and the cytoplasm conferred; *then*, the nucleus told the cytoplasm what to do. The nucleus, like the ideal American husband, still had the power and the final decision; but the decision was made only after discussions with the female partner. Not only was this a more American view of marriage, it was also the relationship between T.H. Morgan and his wife (G. Allen, Personal Communication). A third view came from C.H. Waddington, a British socialist. Waddington married a successful architect and viewed his marriage as a partnership. Werskey (1978, 221) has pointed out that Waddington respected women as intellectual equals, and Waddington viewed the marriage of nucleus and cytoplasm as a partnership. In *Organisers and Genes* (1940), Waddington tried to show the equality of nucleus and cytoplasm, neither dominating the other. His cell, like his notion of marriage, was a partnership between equals. The fourth view comes from the American Black embryologist E.E. Just (1939) who declared the cytoplasm to dominate over the nucleus. The nucleus was subservient to the commands given it by the cytoplasm, and only the cytoplasm was endowed with vitality. This also reflects Just's view of male/female relationships, for "Just saw himself working for Hedwig [his lover] as a slave works for his master" (Manning 1983, 265). For Just, who viewed fertilization largely as a consequence of the cytoplasmic activity of the egg, the male was subservient to the female. Thus, all four views of nuclear/cytoplasmic interactions reflect views of male/female interactions.

Contemporary biology, although aware of the interactions of the cytoplasm and nucleus, still tends to portray the nucleus as the head of the family's hierarchy. Jacob (1976, 224) writes, "Among all the constituents of living organisms, the genetic material has a privileged position. It occupies the summit of the pyramid and decides the properties of the organism. The other constituents are charged with the execution of the decision." The term "genetic engineering" (like "reproductive technology") is a masculine metaphor appropriating the role of procreation to technology. Haraway (1984) claims that "genetic engineering . . . is a science fiction expression suggesting the triumph of the phallogocentric lust to recreate the world without the intermediary of fleshy women's bodies." In genetic engineering, the assumption has been that DNA is the "master molecule," and introductory biology texts still call DNA by that name.[6] This isn't surprising given the hierarchical "central dogma" of DNA→ RNA→ Protein and the views of J.D. Watson ("the best home for a feminist is in another person's lab"). David Nanney (1957, 136) and Evelyn Fox Keller (1985, 150) have criticized this view, and Nanney has put forth an alternative model. He argues against the "Master Molecule concept. . . . This is in essence the theory of the Gene, interpreted to suggest a totalitarian government." He opposed this to "The 'Steady State' concept. By this term . . .we envision a dynamic self-perpetuating organization of a variety of molecular species which owes its specific properties not to the characteristic of any particular molecule, but to the functional interrelationships of these molecular species." E.E. Just, in fact, had criticized

McClung's notion of chromosomal hegemony on the same grounds. McClung (1924, 634) had claimed that, "Taken together, the chromosomes represent the sum total of all the elements of control over the processes of metabolism, irritability, contractibility, reproduction, etc., that are involved in the life of the organism." Note the use of the nucleus as the repository of all the *control* functions of the cell. Just (1936, 305) replied that "Such statements are absolutely without foundation in fact." Just (1936, 292) also linked nuclear hegemony with authoritarianism. It is not surprising that Nanney is one of the leading authorities on extrachromosomal inheritance and the cell cortex, and that E.E. Just attempted to popularize E.B. Wilson's observations on the eggs' activity in fertilization.

The master-molecule has become, in DNA, the unmoved mover of the changing cytoplasm. In this cellular version of the Aristotelian cosmos, the nucleus is the efficient cause (as Aristotle posited the sperm to be) while the cytoplasm (like Aristotle's conception of the female substrate) is merely the *material* cause. The nuclear DNA is the essence of domination and control. Macromolecule as machomolecule. Keller (1985) notes that on the cellular level, the hierarchical depiction of DNA in most textbooks looks like "organizational charts of corporate structures" and that genetic stability is ensured by the unidirectionality of information flow, much as political and social stability is assumed in many quarters to require the unidirectional exercise of authority." This hierarchy on the cellular level is supported by sociobiology on the organismal level. Here, bodies are merely vehicles for the propagation of genes. They are the fruit which nourishes the seeds. Similarly, the metaphors of sociobiology are drawn from the investment economics of our present society (Haraway 1979; Schwartz 1986).

The steady-state view of the cell is presently a minority opinion, but it has recently been eloquently expressed by Lynn Margulis and by Lewis Thomas (1974, 1). Here, the cell is seen as an ecologically interacting entity where process and interrelatedness are fundamental characteristics of life, not the properties of a single molecule.

The modeling of the nucleus began with a template of domination: "What controls what?" This was secondarily sexualized such that the nucleus (male) was seen as dominating the passive (female) cytoplasm. This sexualization of the cell has had enormously important affects on how biologists view the cell and this view, now "objectified" by science, supports the social behaviors which imposed it in the first place. The sexualization of the cell has placed blinders on researchers, making certain observations (and interpretations) "normal" and others "aberrant." In this section, we have tried to show that the tendency to equate activity with masculinity and passivity with femaleness has caused the research programs of fertilization and sex determination to be directed in a way different than it might have otherwise been. But can such degenderization succeed, or are we engrained in our telling of sexual stories? There is a case where the degenderizing of the cell has succeeded to the benefit of the science. In protozoology at the turn of the century, gender distinctions had been placed on unicellular organisms (a strange situation considering these are cells and lack vaginas, penises, ovaries or testes). M. Hartmann (1929), one of the leading protozoologists of his time held that whenever differences were found within species, these differences would be male and female. In an article opposing this view, T.M. Sonneborn (1941, 705) noted that "the characteristics by which the female is ordinarily recognized are larger size, lesser activity, greater storage of nutritive reserves, and egg-like form; and the male by the corresponding opposite characters." Sonneborn pointed out that this dichotomy had created artificial problems that had directed research into less productive areas, and that a better protozoology could emerge if the male and female distinctions were abandoned. Sonneborn's ideas prevailed, and the analysis of mating types (*plus* and *minus*: "a" and "alpha"; not male and female) has become one of the most exciting areas of the field.

FERTILIZATION METAPHORS IN ORGANIC CHEMISTRY

The sperm-egg interaction is a metaphor in-and-of-itself. Sometimes, the metaphor is explicit and sometimes implicit, but many things appear to interact "like the sperm and egg." Implied in this analogy is an active partner and a passive partner. We see this in many introductory textbooks of organic chemistry. Collisions between two molecules which lead to the formation of new compounds are often depicted sexually or aggressively, an active, small molecule "attacking" a large, passive, heavy compound. Nucleophilic and electrophilic "attacks" are standard language in organic chemistry. "The entering group is a negative species which is attacking the nucleus of the reactive carbon . . ." (Cason 1966, 66, 76). In the same book, college sophomores are also taught that "the nucleophile attempting a backside attack on the molecule is confronted with a problem that may be likened to the effort to penetrate a set of propellers spinning at high speed."

The notions of penetration and entry are often standard parts of organic chemistry lectures. It is not surprising to read that the "characteristic reaction of a carbene is insertion." Another book (Cook and Crump 1969, 71) describes the alkene bond as "being 'ripe for plucking' by an approaching electrophile." The heroic nucleophile or electrophile must be, like the sperm, tested. "The potency of a nucleophile in affecting a displacement is termed its nucleophilicity or nucleophilic strength" (Cason 1966, 363).

Who would have expected nucleophallic and electrophallic molecules? It appears that an arbitrary genderization of molecules has been made, where one of the colliding molecules is called the "attacking" group and the other is the passive recipient of this attack. In both nucleophilic and electrophilic "attack," the "attacking" molecule is not the larger, but the smaller, faster one. The large molecules, those that are "looser" in terms of their electronic configuration (more resonance, pi-bonding) are the passive attacked groups. This arbitrary imagery is, we believe, analogous to small, hard mobile sperm penetrating the large, soft, immobile eggs. The imagery conforms to stereotypic attributions of maleness to energetic elements and femaleness to the passive ones. These stereotypes are being propagated by the language of science which gives students a wrong idea of nature (i.e., that it is gender-biased) but which purports to be objective.

NATURE AS TEXT

> "Like other sciences, biology today has lost many of its illusions. It is no longer seeking the truth. It is building its own truth."
> —Francois Jacob (1976,16)

Science is a creative human endeavor whereby individuals and groups of individuals collect data about the natural world and try to make sense of them. Each of the basic elements of scientific research—conceptualization, execution and interpretation—involves creativity. In fact, these three elements are the same as most any artistic, literary or musical endeavor. Two aspects of science are especially creative, namely the conceptual designing of an experiment and the interpreting of the results. Usually, the interpretation is put in the context of a narrative which includes the data but is not dependent upon them (Medawar 1963, 377; Figlio 1976, 17; Landau 1984, 262). Since science is a creative endeavor, it should be able to be criticized as such; and Lewis Thomas (1984, 155) has even suggested that schools of science criticism should exist parallel to that of literary, music and art criticism.

As a creative part of our social structure, biology should be amenable to analysis by feminist critique which has provided new insights into literature, art and the social sciences. Indeed, feminist examinations of sociobiology (Sayers 1982; Bleier 1984) primate research (Haraway 1986), and scientific methods (Keller 1985) have provided an important contribution to the literature of those fields. Researchers in those fields are aware of the feminist criticism and the result has created a better science—one in which methods of data collection and interpretation have been scrutinized for sexual biases.

Any creative enterprise undertaken by human beings is subject to the influences of society. It is not surprising, then, to see how gender becomes affixed to cells, nuclei and even chemicals. Even the interpretations of mathematical equations change with time! The interpretation that Newton gave to his Law of Gravity (i.e., that it was evidence of God's power and benevolence) differs (Dobbs 1985) from the interpretation of eighteenth century physicists (that it was evidence for a mechanical universe devoid of purpose), and from that of contemporary physicists (that it is the consequence of gravitons traversing the curvature of space around matter).

By using feminist critique to analyze some of the history of biological thought, we are able to recognize areas where gender bias has informed how we think as biologists. In controlling for this bias, we can make biology a better discipline. Moreover, it is important that biology be kept strong and as free from gender bias as possible; for it is in a unique position to do harm or good. As Heschel has remarked (albeit with masculine pronouns):

> The truth of a theory about man is either creative or irrelevant, but never merely descriptive. A theory about the stars never becomes a part of the being of the stars. A theory about man enters his consciousness, determines his self-understanding, and modifies his very existence. The image of a man affects the nature of man . . . We become what we think of ourselves. (1965, 7)

A theory about life affects life. We become what biology tells us is the truth about life. Therefore, feminist critique of biology is not only good for biology but for our society as well. Biology needs it both for itself and for fulfilling its social responsibilities.

NOTES

*We wish to thank Donna Haraway, Evelyn Fox Keller, Sharon Kingsland, Jeanne Marecek and Nancy Tuana for their comments on earlier drafts of this paper.

Lest anyone believe that this is strictly an academic exercise, the *New York Times* (25 March 1987, Sec. I, p. 20) recently reported an article wherein Adrianus Cardinal Simonis, Primate of the Netherlands, cited fertilization as evidence for the passive duties of women. In this essay, the Archbishop pointed to the egg that merely "waits" for the male's sperm, which he described as the "dynamic, active, masculine vector of new life."

1. The apparent exception of mammalian males was considered due to the extra burden *they* had when their mates were pregnant.

2. Once given "objectivity" by science, the notion that men are active because of their spermatic metabolism and women are passive because of their ovum-like ways finds its way into popular definition of masculinity and femininity. Freud (1933, 175) felt it necessary to counter this view when he lectured on "Femininity": "The male sex-cell is actively mobile and searches out the female one, and the latter, the ovum, is immobile and waits passively . . . The male pursues the female for the purpose of sexual union, seizes hold of her and penetrates into her. But by this you have precisely reduced the characteristsic of masculinity to the factor of aggressiveness as far as psychology is concerned." Freud recognized that "it is inadequate to make masculine behavior coincide with activity and feminine with passivity," and that "it serves no useful purpose and adds nothing to your knowledge."

3. There is ample evidence for the ovum as mythic princess. The ovum is not allowed to see sperm before it is of age, and when it travels to meet the sperm this "ripe" ovum not only has a "corona" (crown) but "vestments." It is also often said to have "attendant cells." According to Jung (1967, 171, 204), the hero is the symbol *par excellence* of the male libido and of the longing to reunite with the mother. If true, the sperm is an excellent embodiment of the heroic fantasy.

But this does not mean we have to follow this myth. Indeed, one could make a heroic tale about the ovum which has to take a "leap" into the unknown, though its chances of survival are less than 1%. Indeed, the human ovum, too, is a survivor of a process which has winnowed out nearly all of the original 2 million oocytes, and left it the only survivor of its cohort.

4. The "burrowing" metaphor is also commonly seen in textbooks, and it brings with it the seed-and-soil imagery. This plowing trope was, for many ancient cultures, a metaphor of necessary violence. The active/passive dichotomy is remarkably evident in the verb *to fertilize*. The traditional statement is that the "sperm fertilizes the egg." The sperm is active, the egg is passive. This inverts the original meaning of *fertilize* which involves the nourishment of seeds *by* the soil. The verb no longer connotes nutrition in this context, but activation.

5. Although Eicher and Washburn have emphasized that both sexes are actively created, at least two reviews on sex determination have recently proposed one or the other sex as being the "default" condition of the species. It should be noted that the views expressed in this essay may or may not be those of the scientists whose work we have reviewed. It is our contention that these research programs are inherently critical of a masculinist assumption with these respective fields. This does not mean that the research was consciously done with this in mind.

6. Metaphor and connotive language is extremely important in producing the gender-related images. Introductory biology textbooks also refer to the pituitary as "the master gland." (After all, it controls the other organs of sex and internal secretion from its privileged position in the brain. The apical, brainy organ controls the organs of lower functions; the sex glands being furthest removed.) There are other metaphors that could have been utilized. The pituitary could be called the "switchboard" gland (a female gender image) or the "integrator" gland (a dialectical image). Similarly, it is not merely a figure of speech to say that the seed analogy is at the heart of cell biology. The German word *Kern* (and Germany was where most of the pioneering work on cytology and fertilization was done) means more than the English equivalent "nucleus." It also means kernel, center, quintessence and elite position. Similarly both sperm and semen (and their German equivalents) have the same etymology, namely "seed." *Mater*, however, gives the root for maternal, material, matter and matrix.

The seed metaphor was so real to Leeuwenhoek that he actually performed dissections of plant seeds, insisting that the embryonic human would be found in the sperm just as the embryonic plants were found in the seeds (Ruestow 1983, 204). His "spermatozoa" were precisely that: mobile, ensouled, seed-animals. To him, the uterus (and the female sex) served to nourish the seed. The father was the sole parent.

REFERENCES

Bleier, R. 1984. *Science and gender: A critique of biology and its theories on women.* New York: Pergamon Press.

———. 1986. *Feminist approaches to science.* New York: Pergamon Press.

Boylan, M. 1984. The Galenic and Hippocratic challenges to Aristotle's conception theory. *Journal of the History of Biology* 17:83-112.

Campbell, J. 1956. *The hero with a thousand faces.* Cleveland: Meridian Books.

Cason, J. 1966. *Principles of modern organic chemistry.* New Jersey: Prentice-Hall.

Cook, P.L. and J.W. Crump. 1969. *Organic chemistry: A contemporary view.* Lexington, MA: Heath.

Dobbs, B.J.T. 1985. Newton and stoicism. *Southern Journal of Philosophy* 23 (Supp):109-123.

Eicher, E.M. and L. Washburn. 1986. Genetic control of primary sex determination in mice. *Annual Review of Genetics* 20:327-360.

Fausto-Sterling, A. 1985. *Myths and gender: Biological theories about men and women*. New York: Basic Books.

Figlio, L.M. 1976. The metaphor of organization. *Journal of the History of Science* 14:12-53.

Freud, S. [1933] 1974. Femininity. In *Women in analysis*, ed. J. Strouse. New York: Grossman.

Geddes, P. and J.A. Thomson. 1890. *Evolution and sex*. New York: Moffitt.

———. 1914. *Problems of sex*. New York: Moffitt.

Gilbert, S.F. 1985. *Developmental biology*. Sunderland, MA: Sinauer Associates.

———. In Press. Cellular politics: Goldschmidt, Just, and the attempt to reconcile embryology and genetics. In *The American development of biology*, ed. K. Benson, J. Maienschein and R. Rainger. University of Pennsylvania Press.

Haraway, D. 1979. The biological enterprise: Sex, mind, and profit from human engineering to sociobiology. *Radical History Review* 20:206-237.

———. 1984. Lieber Kyborg als Gottin! Fur eine sozialistische—feministische Unterwanderung der Gentechnologie. In *Argument-Sonderband* 105, ed. B.P. Lange and A.M. Stuby, 66-84.

———. 1986. Primatology is politics by other means. In *Feminist approaches to science*, ed. R. Bleier, 77-119. New York: Pergamon Press.

Hartmann, M. 1929. Verteilung, Bestimmung, und Vererbung des Geschlechtes bei den Protisten und Thallophyten. *Handb. d. Verer*, II.

Harwood, J. 1984. The reception of Morgan's chromosome theory in Germany: Inter-war debate over cytoplasmic inheritance. *Medical History Journal* 19:3-32.

Heschel. A.J. 1965. *Who is man?* Stanford: Stanford University Press.

Horowitz, M.C. 1976. Aristotle and woman. *Journal of the History of Biology* 9:183-213.

Jacob. F. 1976. *The Logic of life*. New York: Vintage.

Jung, C.G. 1967. *Symbols of transformation*. Princeton: Princeton University Press.

Just, E.E. 1936. A single theory for the physiology of development and genetics. *American Naturalist* 70:267-312.

———. 1939. *The biology of the cell surface*. Philadelphia: Blakiston.

Keeton, W.C. 1976. *Biological science*, 3rd ed. New York: W.W. Norton.

Keller, E.F. 1985. *Reflections on gender and science*. New Haven: Yale University Press.

Landau, M. 1984. The narrative structure of anthropology. *American Scientist* 72:262-268.

Lenoir, T. 1982. *The strategy of life*. Dordrecht: D. Reidel.

Manning, K.R. 1983. *The black apollo of science: The life of Ernest Everett Just*. New York: Oxford University Press.

McClung, C.E. 1901. Notes on the accessory chromosome. *Anatomischer Anzeiger* 20.
———. 1902. The accessory chromosome—Sex determinant? *The Biological Bulletin* 3.
———. 1924. The chromosome theory of heredity. In *General Cytology*. Chicago: University of Chicago Press.
Morgan, T.H. 1926. *The theory of the gene*.. New Haven: Yale University Press.
Medawar, P.B. 1963. Is the scientific paper a fraud? *The Listener* (12 September): 377.
Nanney, D.L. 1957. The role of the cytoplasm is heredity. In *The chemical basis of heredity*, ed W.E. McElroy and H.B. Glenn, 134-166. Baltimore: Johns Hopkins University Press.
Reil, J.C. 1795. Von der Lebenskraft, *Arch. f.d. Physiol.* 1. Quoted in *The strategy of life*. See Lenoir 1982.
Ruestow, E.G. 1983. Images and ideas: Leewuenhoek's perception of the spermatozoa. *Journal of the History of Biology* 16:185-224.
Russell, K.P. 1977. *Eastman's expectant motherhood*. 6th ed. New York: Little.
Sayers, J. 1982. *Biological politics: Feminist and anti-feminist perspectives*. New York and London: Tavistock.
Schatten, G. and H. Schatten. 1983. The energetic egg. *The Sciences* 23 (5):28-34.
Schwartz, B. 1986. The battle for human nature: Science, morality, and modern life. New York: W.W. Norton.
Sonneborn, T.M. 1941. Sexuality in unicellular organisms. In *Protozoa in biological research*, ed. G.N. Calkins and F.M. Summers. Chicago: University of Chicago Press.
Thomas, L. 1974. *The lives of a cell*. New York: Viking.
———. 1984. *Late night thoughts on listening to Mahler's ninth symphony*. New York: Bantam.
Werskey, G. 1978. *The visible college*. New York: Holt, Reinhart, and Winston.
Waddington, C.H. 1940. *Organisers and genes*. Cambridge: Cambridge University Press.

The Premenstrual Syndrome "Dis-easing" the Female Cycle

JACQUELYN N. ZITA

This paper reflects on masculinist biases affecting scientific research on the Premenstrual Syndrome (PMS). Masculinist bias is examined on the level of observation language and in the choice of explanatory frameworks. Such bias is found to be further reinforced by the social construction of "the clinical body" as an object of medical interrogation. Some of the political implications of the medicalization of women's premenstrual changes are also discussed.

Recent interest in Premenstrual Syndrome has been carried on a wave of ideological panic. Premenstrual syndrome has been accepted as "a disease of mind" in English courts of law (Luckhaus 1985; Allen 1984; Brahams 1981), and for two women this has meant that charges of murder have been commuted to manslaughter by reason of diminished responsibility due to PMS. This led Dr. Gerald Swyer (Nicholson 1982, 227) to testify in one of these court cases that a PMS defense will legitimate excuses for all kinds of irrational female behavior — women may now kill, batter and aggress with impunity. At the same time, new studies have emerged on the cost of PMS to employers of industrial plants (Dalton 1979; Parker 1960), costs spiralling into billions of dollars due to absenteeism and sedation.

Correlational studies have been used to confirm this panic. These studies have shown women to be more prone to violence and irritability during the premenstrum (Ivey and Bardwick 1968), more likely to have accidents (Dalton 1960; MacKinnon and MacKinnon 1956), more likely to have morbid fears and sexual fantasies (Benedek 1959), more likely to commit suicide (Tonks 1968; Mandell 1967), more likely to take their children to doctors (Dalton 1966), more likely to break the law (Dalton 1961), more likely to skip work and seek sedation (Dalton 1979), more likely to batter children (Dalton 1977), more likely to abuse alcohol (Janowsky 1969), and finally more likely to end marriages (Dalton 1979) and crash airplanes (Whitehead 1934).

Katharina Dalton in her book, *The Premenstrual Syndrome and Progesterone Therapy* (1977, 150-157), has even suggested that the premenstrual mother is responsible for the diminished performance levels of all family members:

Hypatia vol. 3, no. 1 (Spring 1988). © by Jacquelyn N. Zita.

the schoolchild, the teenager, grandparents and husband. She writes, "One survey showed that the husband's late arrival at work was a reflection of the time of his wife's cycle. They both failed to get up with the alarm, they quarrelled over breakfast which consequently took longer, and then the sandwiches weren't ready!" (1977, 156). Not only is the premenstrual woman a risk factor in the workforce; she has now become a hazard in the home!

Thus, the concept of PMS has been recently used to describe deviance in female behavior, a deviance which is sensationalized in the media. These behaviors are extreme—murder, battery, suicide, accidents, violence—and seem to be related to women's cyclicity. They have also been labeled as part of a "syndrome," with the concomitant physiological and psychological changes categorized as "symptoms." The language of "syndrome" and "symptom" suggests that at least some women's social behavior and psychophysiological changes could be construed as a clinical entity.[1] So codified, Premenstrual Syndrome, like all medical syndromes, presupposes the existence of a disease-state with unclear etiology. Prior to this, evidence supporting the existence of behavioral and physiological cyclicity in females had been well established. Cyclicity had been observed in almost every physiological system in the body (Southam 1965; Sommer 1978). Notable among these studies is the classic (1937) McCance study of 167 healthy women who kept daily self-report records for a total of 780 cycles. His research team found periodicity with respect to certain variables, such as breast changes, incidence of depression, tendency to cry, incidence of irritability and others. Increases in incidence rates in these variables and other changes were more or less located within the premenstrual and menstrual phase of the cycle. More recently, this recurrent pattern was defined in 1977 by Katherina Dalton as "the recurrence of symptoms on or after ovulation, increasing during the premenstrum and subsiding during the menstruation, with complete absence of symptoms from the end of menstruation to ovulation." Over the past decade this definition has been maintained as seen in the 1985 definition suggested by the National Mental Health Association: "a constellation of mood, behavior and/or physical symptoms which have a regular cyclical relationship to the luteal phase of the menstrual cycle, are present in most (if not all cycles), and remit at the end of menstrual flow with a symptom-free interval of at least one week each cycle" (Harrison 1985, 789). It should be noted that the step from observable cyclicity to the presumption of pathology, which requires disease-model thinking, is an epistemic leap that requires careful scrutiny of the evidence.[2]

In this paper I will re-examine the medical evidence that has been used to posit the existence of PMS as a new clinical entity and the causal explanations that have been proffered to account for this syndrome. My analysis will show that much of the research being done on PMS is, in fact, suspect and that the stature and popularity of the idea of PMS rests upon invidious ideological assumptions about the nature of woman. I will analyze these assumptions on two different levels: those which operate on the level of observational practice and those which are built into explanatory frameworks. My objective is to show how these biases in observation and explanation contribute towards linear reductionist thinking in medical research with its assumed entitlement to medical management of the female body.

THE OBSERVATION PROBLEM

In this section I will examine how the observational language of PMS research is riddled with various assumptions or biases which facilitate disease-model theorizing. Although these assumptions may not be consciously used in a conspiratorial fashion, they are ideological lines of influence which affect the quality and veracity of observational language upon which scientific theorizing is based. The bias in these assumptions affects how observations are made, how and why questions are asked, how observations are used as evidence for the existence of a syndrome, which observations are considered significant or insignificant, and other aspects of experimental design. Furthermore, this bias is best characterized as masculinist, where "masculinism" is defined as an ideological perspective in which gender differences are depicted as binary oppositions, negatively weighted in favor of males and used to justify male domination over women. PMS data concerning women often constitutes just such a collection of negative facts about women's nature, a nature which in turn is seen as requiring medical surveillance and management, along with a "protective" secondary citizenship.

(A) THE NEGATIVITY ASSUMPTION

Within the popular literature on Premenstrual Syndrome, it is often and most clearly a masculinist perspective from which observations are made and advice given. Women's subjective reports are quickly codified as symptoms, by a masculinist voice which assumes an objective authority. This can be seen in the popular works of Katharina Dalton, who oddly enough uses two masculinist voices — the voice of medical "objectivity" which systematically belittles and ridicules female behaviors and the subjective voice of the male victim who suffers these female behaviors:

> Most men enter marriage sublimely ignorant of the problems women face each month; such knowledge as they have is probably confined to a vague awareness of the "monthlies" when she bleeds and is "on the rag." Before marriage and whilst they were going together, it was easy enough for the girl to conceal those difficult days and all too often it is not until they

> are living together that the awful truth begins to dawn. If she is a sufferer of spasmodic dysmenorrhea he will be the first to see that she gets good and complete relief for her pains, for pain is something a man can understand. However, sudden mood changes, irrational behavior, and bursting into tears for no apparent reason are bewildering, while sudden aggression and violence are deeply disturbing when with little warning and no justification, his darling little love bird becomes an angry, argumentative, shouting, abusive, bitch. (1977, 79-80)

Dalton urges a husband to keep a monthly diary of his wife's behaviors, and if he is a sensible husband, he will soon recognize the time this occurs in the relationship and, "being sensible, will insist on her seeing a doctor and obtaining treatment for her premenstrual syndrome" (1977). While on the surface this cajoling paternalistic interest in wives seems loving, the subtexts present women as wildly out of control, incapable of sensible judgement, and dependent on their husbands and doctors for help (1979, 83). Dalton further adopts a masculinist perspective in blaming marital discord and family stresses on these periodic lapses in female placidity.

> Sometimes the marital discord is just silence, on other occasions there may be vicious verbal battles, and at the extreme limits there are fights and batterings. How many wives batter their husbands during their premenstrum is unknown, nor do we know how often the husband is provoked beyond endurance and batters her. (1979, 85)

Although Dalton's misogynistic assumptions can be easily recognized, this does not mean that she is an exception in the medical literature. Wickham (1958), for example, after describing menstruation as a "physical disturbance," continues to write descriptively, "towards the end of the cycle, the *destructive forces are in control* and they culminate in the menstrual period" (emphasis added). The observation language used and the interest behind much PMS research are directed towards the description and depiction of negatively evaluated changes in women. Parlee (1974) has pointed out that the Moos Menstrual Distress Questionnaire, a common PMS research tool, is almost entirely focused on negative conditions. As might be expected, much less attention is given to the alleged midcycle peak in the female cycle (Altman 1941; Ivey 1968), which is for many women a time of euphoria, creativity and heightened productivity. Positive aspects of the premenstrum are commonly ignored. Instead attention is almost completely riveted on "what goes wrong with women" once a month.

(B) THE BASELINE PROBLEM

Although it may be true that some women believe something does "go wrong" once a month, medical fixation on this negativity makes easier the transition from simple descriptive language to disease-model theorizing. "Disease" is defined as an undesirable deviation from the norms of physical and psychological well- functioning (Boorse 1975). The concept of deviance requires a measure of distance from some norm. Within the literature on PMS, there are two such implicit norms: male physiology and feminine behaviors. Against the first norm, female cyclicity is considered deviant when compared to relatively non-cyclic male physiology. In short, cyclicity is problematized. Cyclicity is further troubled when measured against the norms of continuously placid femininity. The conditions associated with the premenstrum, such as irritability, aggression, violence, moodiness, are seen as "deviant" (if "deviant" is defined as deviant from the norms of passive femininity), and "undesirable" (if this is what is not desirable in women) and apparently incapacitating (if "disability" is defined as not accepting feminine roles or as disruptive to the linear organization of male social labor). Instead of seeing these "deviations" as measured against questionable social norms of benign femininity and masculine "linearity," they are translated into deviance from the norms of physical and mental well-functioning, that is, health. The collected data is thus transmuted into the language of "symptoms" and "signs" within disease-model theorizing. Dalton (1977) crystalizes this in her description of one patient, who after progesterone treatments, "felt feminine again."

Furthermore, the norms against which deviance is measured are often presupposed without careful comparison. As Parlee (1973) has pointed out, control groups consisting of nonmenstruating individuals are seldom used in PMS research. The absence of such controls biases the explanatory theorizing, since it is simply assumed that it is something female, usually female hormones, which causes the syndrome. Such a presumption should be tested against a control group of males, but this is seldom done, even though some research suggests that men also undergo marked patterns of cyclicity (Kihlstroem 1971; Doering 1975; Lieber 1972). If such patterns are linked with male hormonal cycles, female steroid cyclicity becomes a matter of degree rather than a qualitative or absolute sex difference. If male cyclicity is linked to non-hormonal factors, the explanation for female cyclicity may have to introduce these factors as well. If, on the other hand, a control group consisted of women not suffering from PMS, the causal factors explaining PMS might have to include more than a simple reductionist hormonal explanation, since some recent research suggests that there is no difference between the hormonal profiles of women suffering from PMS and those not so afflicted (Sondheimer 1985).

The omission of careful control studies results in an over- rating of PMS symptoms. Consider the correlational evidence Dalton (1960) presents between the rate of accidents and the time they occur in women's menstrual cycles. In this study, she tabulates the distribution of 84 accidents in which 84 regularly menstruating women were involved. Disease-model thinking requires a descriptive set of correlations that can be related to underlying causalities. Dalton's results suggest a strong correlation between higher accident rates and the late premenstrum and/or menstruation. This deviation from the expected rate is taken to be significant, although Dalton in her analysis never establishes statistical significance. What is further omitted from her study is any comparison between female and male accident rates, since such a comparison would clearly show that female accident rates are far below that of the opposite gender. Without this comparison, the deviation from the expected "feminine" rates is over-rated, as is the danger assumed regarding female drivers. In addition, Dalton's study includes only those women who have had accidents, although she uses this as evidence to describe the accident proneness of all women during the premenstrum. This is further overdrawn by her elimination of other factors that could trigger menstruation. As Sommer (1978) and Parlee (1973) have noted, accidents may precipitate menstruation, thus increasing the apparent premenstrual accident rate among women. This would account for the high incidence of onset of menstruation among women who were the "passive participants" (i.e., vehicle passengers) in these studies (Dalton 1960, 1425).

The baseline problem is that of defining the norms or the reference points against which deviation is measured and evaluated. In many research articles statistical significance is not measured, the assumption being that increased incidence rates in conditions at or near the time of the premenstrum require no further analysis. Even when statistical significance is tabulated, such a measure of difference does not necessarily entail a clinical indication of morbidity, since deviation from a given norm may well be within the bounds of physiological normalcy. What is lacking in some PMS research is a careful measure of the severity of premenstral changes,[3] a measure of the incapacitating effects associated with premenstrual changes and a careful scrutiny of behaviors and norms against which "incapacitation" is measured and evaluated. To assume that any deviation is abnormal, disabling and deserving of medical attention begs the question of disease-model thinking. As in the case of dysmennorhea, many women suffer from menstrual cramps, but only 8-13% (Sommer 1982) suffer from severe and disabling dysmennorhea. This figure has not led the medical community to promote dysmennorhea as a universal female disease, although some severe cases may be indicative of various underlying pathologies, not necessarily a singular disease-state. The same may be true of severe and disabling premenstrual changes.

(C) SYMPTOMOLOGIES

Describing and defining which symptoms are associated with the premenstrual syndrome has always been problematic in this research. Janowsky and his co-workers established the following list of cyclically recurring phenomena of the premenstrual and menstrual phases:

> depression, irritability, sleep disturbances, lethargy, alcoholic excesses, nymphomania, feelings of unreality, sleep disturbances, epilepsy, vertigo, syncope, nausea, vomiting, constipation, bloating, edema, colicky pain, enuresis, urinary retention, increased capillary fragility, glaucoma, migraine headaches, relapses of meningiomas, schizophrenic reactions and relapses, increased susceptibility to infection, suicide attempts, admission to surgical and medical wards, crime rates, work morbidity, manic reactions, and dermatological diseases. (1966, 54)

Moos (1968) cited over 150 symptoms that characterized the premenstrual syndrome. These were organized into five different categories and their subcategories:

> *Affective*: sadness, anxiety, anger, irritability, labile mood.
> *Neurovegetative*: insomnia, hypersomnia, anorexia, craving of certain foods, fatigue, lethargy, agitation, libido change.
> *CNS*: clumsiness, seizures, dizziness, vertigo, tremors.
> *Cognitive*: decreased concentration, indecision, paranoia, "rejection sensitive," suicidal ideation.
> *Behavioral*: decreased motivation, poor impulse control, decreased efficiency, social isolation.

There are other lists of symptoms and other ways of categorizing the symptoms that have appeared in the literature, all of which creates a most unusual medical "syndrome," one that is not defined by a clearly established and consistent set of symptoms but by the timing of when any of a number of different symptoms occur during the menstrual cycle. Further problems emerge when consistency is sought, since different measuring techniques produce different symptoms which are often not commensurable or comparable. As was mentioned earlier, the measures used are at times vague and qualitative, with very few indications of severity or even differences within a singular category.

Given this broad and inconsistent listing of "symptoms," it is not surprising to find exceedingly high prevalence rates. Sutherland and Stewart (1965), given their definition of PMS, found that only 3% of their sample of healthy women could be said not to suffer from PMS. A 95% prevalence rate is frequently found in the literature (Pennington 1957; Seagull 1974), suggesting that these conditions afflict almost the entire female population. When the severity of conditions is studied, however, the prevalence rates for severity are considerably lower. Even Dalton has suggested that only 10% of her special self-selected client population can be said to have severe and incapacitating premenstrual conditions. In spite of this, it is a popular assumption that every woman runs a high chance of experiencing the "syndrome." This is supported by the variable grid of symptoms associated with the syndrome, since almost any physical or psychological change can be countenanced as long as it occurs during the premenstrum.

(D) GATHERING THE EVIDENCE

All of this is further compounded by complications encountered in the collection of data used to confirm the existence of this alleged syndrome. Quite often studies are based on clinical populations of women attending premenstrual treatment clinics or infertility clinics or women who have come to their doctors with premenstrual complaints. Such research will be biased by its population base, since these are women who have already self-selected themselves as having a problem and needing help. Research based on wider community populations is costly and also burdened with further methodological problems in data collection.

McCance (1937), Parlee (1982b), and Abplanalp (1979) have all noted discrepancies between subjects' retrospective reports when using questionnaires (or interviews) and their ratings in concurrent studies using self-report techniques. The conclusion is that subjects tend to overestimate their conditions when working from memory, since methods such as daily self- reports give a much less dramatic description of the subject's periodicity. PMS research that is based on retrospective data collection may be measuring the subjects' attitudes and assumptions about premenstrual conditions rather than what is actually experienced.

Social expectations may also play an intervening role. Women's aggressive behavior during the premenstrum may reflect what others expect from women and women from themselves during that period. Given this social attitude, women may give themselves more permission to be irritable, angry or aggressive. Others, likewise, come to expect it. As a result what is being measured in subjective reports is greatly influenced by expectations from others and preconceived notions about what PMS does to women. I will return to this point later.

(E) FROM EVERYDAY COMPLAINTS TO CLINICAL LANGUAGE

The ideological meaning of gender plays an important role in establishing the ground-level observational language used in PMS studies. In this research, there are two different modes of observational language: (A) informal observation based on male and female subjective interpretations of changes in female bodily or behavioral functions, usually registered as complaints , and (B) more formal observations made in clinical or experimental settings where these subjective reports are codified into quantifiable symptoms and signs sharing in common a pattern of premenstrual cyclicity. These levels of observational language can be schematized as follows:

PRIMARY DATUM
The Subject's Verbal/Non-Verbal Behavior

OBSERVATIONS A
(COMPLAINTS) Everday Discourse

(A1) The Subject's Informal Reports About Herself
(A2) Others' Informal Subjective Reports About the Subject

OBSERVATIONS B
(SYMPTOMOLOGIES)
Clinical Discourse

(B1) Quantifiable Physiological Changes
(B2) Quantifiable Psychological Changes
(B3) Quantifiable Behavioral Changes

Observations of Type (A) often incorporate negative ideological assumptions about the female gender, such as common attitudes regarding women's presumed irrationality, moodiness, emotional intensity, hysterical behavior, lack of control, etc. This masculinist bias tends to exaggerate gender differences and to evaluate these differences as negatively disfavoring to women. Thus, the language that ascribes negative deviance or difference to women's everyday behavior becomes the first level of descriptive discourse in PMS theorizing. It motivates the need for medical re-description.

Medical discourse, however, is often further removed from women's original subjective expression. For example, a woman who premenstrually displays anger or irritability (Primary Datum) is expressing a meaningful behavior, one that conveys her feelings, perceptions and attitudes towards a particular situation. Once categorized as a negative behavior—"not acting herself today," "irrational," "out of control," "crazy,"—the original behavior has been evaluated against a norm of expected social behavior. This observation (Type A) is then further codified by medical discourse as one of a set of symptoms and signs (observations of Type B), and as a deviation from the norms of well-functioning. Within this discourse the specific meanings of many women's sufferings and anger are homogenized and reified into generic categories (symptoms and signs) which point towards an underlying yet indeterminate pathology. Since the symptomologies used to net this linguistic coup are widely ranging and variously measured, any cyclicity in female behaviors—emotional or otherwise—may be absorbed by this schema. Once generalized over a wide variety of female behaviors, the subjectively expressed meanings voiced by women are replaced by the language of objective (i.e., measurable) signs and symptoms. The intended meanings of women's everyday behavior is thus understood via causal analysis, as women's words and actions come to indicate an indeterminant malfunctioning. The question which requires further consideration is whether the initial naming of difference and devaluation as registered by ideologically-laden complaints (Type A) warrants generalized reification, which strips women's voices from situational contexts and replaces the intended meaning of their behavior with causal categories of explanation. (Type B)?

What I have argued in this section is that the descriptive observational language employed in PMS research tends to beg the question of the appropriateness of the disease-model theorizing. All too quickly women's cyclicity is turned into adversity and reconstrued within the labeling process of "symptom," "syndrome," and "disease." This linguistic *tour de force* is preconditioned by hidden research assumptions which presume the negativity of premenstrual changes, leave unclear the baseline against which deviation is measured, and expand the list of symptoms so that the syndrome seems to become *a fact about women or women's nature.*

It is precisely this that is problematic, since the dominant ideology also purports to give us "facts" about woman's nature based on the notion of female body as obstruction. PMS theorizing mirrors these commonly held social assumptions about women, portraying women as victims of raging hormones and subject to irrational, irritable and unladylike fluctuations in behavior and attitudes. The merging of these ideas into the machinations of scientific discourse presents us with a new "objective legitimation" of woman's nature, a nature that is inherently flawed or at least cyclically "diseased."

PMS: THE EXPLANATORY PROBLEM

In my analysis of PMS observational language, I have already described how many of these observations implicitly presuppose a disease-model explanation. This attitude towards the female body is not novel, since a similar process can be seen in medical approaches to birthing (Arms 1975; Haire 1974), menopause (MacPherson 1981), obesity (Ritenbaugh 1982), battery (Stark 1979), and alcoholism (Ettorre 1986). In all of these cases, the patient's problem, characterized as "an aggregate of symptoms and signs," is considered the appropriate medical object, with the practitioner or researcher constantly reorganizing these aggregrates in accordance with the restraints of various medical models. The trend in these explanatory models is towards reductionism, locating causal variables within the individual or within correlative pathophysiological events. Micro-level physiological deviation is used to explain macro-level deviancy in behavior or reported symptoms. Thus, the trajectory of medical gaze is from the individual into the body, in search of micro-level physiological causes which can sufficiently explain observed phenomena, thus minimizing the impact of socio-cognitive and ideological variables on the biology of the female life cycle and on the lived experience of that biology.

In the research paradigm for PMS this reductionistic approach is often the rule. Medical research is usually limited to two explanatory models: the organic model and the psychogenic model. Organic models[4] range from explanations based on vitamin B deficiency, hypoglycemia, fluid retention caused by dysfunction along the renin-angiotensin-aldosterone axis, elevated prolactin levels, progesterone deficiency, elevated estrogen/progesterone ratios, increased prostaglandins and endogenous opiates. Almost all of these explanatory frameworks assume some connection between internal dysfunction and hormonal dysfunction. Likewise the psychogenic models assume unresolved Oedipal conflicts (Israel 1938), marital discord (Glass 1971), guilt feelings over sexual temptation (Fortin 1958), poor self-experience of menses (Paulson 1961), poor performance of feminine psycho-social roles (Suarez-Murias 1953), high neuroticism scores (Coppen 1963) and other psychogenic factors as underlying causes of neuroendocrine disorders.

As a rule, these psychogenic models are not purely psychological or psychoanalytic, since cyclicity in PMS symptom formation requires reference to hormonal cyclicity. The inadequacy of a purely psychogenic explanation is typified by Viet (1955), who as an early researcher offers a highly speculative psychoanalytic explanation of premenstrual tension:

> Tensions frequently accompany unconscious ruminations that represent re-enactments of prepuberty fantasies. Preadolescence is the stage of defiance and independent urges. Aware of their

> limitations and feelings of insecurity, the young people are perplexed when subjected to aggressive promptings. Premenstrual symptoms are often an expression of the futility of independence they feel and the anxiety associated with further development of physiological forces. It is often met in psychiatric conditions, especially in hysterical personalities, and is usually related to masturbatory guilt. (1955, 599)

Such an explanation may express hidden unconscious forces causing premenstrual behaviors, but these forces do not explain the cyclicity of symptom formation. Typical of most psychogenic explanations, Viet includes another set of pathophysiological factors—"an endocrine background"—as a necessary and conjunctive causal factor in explaining premenstrual repetition of these symptoms. Both kinds of factors—psychological and physiological—are together construed as sufficient explanation for the observed phenomena, although the causal relations between them are left obscure:

> The dynamics of premenstrual depression involve conflicts about the feminine role and subsequent feelings of inferiority, frustration, disappointment, aggression and guilt. The well-adjusted girl may have conflicts about accepting the feminine role, but she is able to cope with them. This suggests that the psychological constellation alone cannot explain premenstrual syndrome. One cannot infer from certain unconscious conflicts a disease entirely. These conflicts may be present in persons without any pathological symptomology. Premenstrual depressions have an endocrine background. (1955, 600)

These two models for PMS research have expanded into a multitude of competing etiologies, none of which seems to adequately explain all manifestations of the premenstrual syndrome. This appearance of research as usual, with its dispersion of competing micro-reductionist etiologies and multiplication of catalogued symptoms, gives the semblance of science as usual. Science as usual is a legitimating activity driven by questions that are never considered a contrivance. However, part of what we may be witnessing is the forced entry of certain ideological assumptions into the machinery of science or the use of scientific theorizing as a legitimation strategy for certain ideological assumptions. This would give an alternative reading to why there is such a rapid multiplication of competing etiologies, constant shifting of symptomologies and conflicting definitions of the syndrome: the wide variability of symptoms and purported pathologies indicates how variably cyclicity is experienced by women. However, it is assumed that all of this can be subsumed under the concept "premenstrual syndrome." The ideological assumption is that there is one syndrome, universal to and variably present in all women, and explained preferably by one or several etiologies. This one-syndrome assumption is related to the medical construction of PMS as a female gender trait.

This would also explain why certain questions are not being asked by many researchers. Why, for example, is it assumed that there is only one syndrome, unless once again there is an interest in reinstating the syndrome as a fact about female nature? Why not assume that women with severe and disabling symptoms aggravated premenstrually may have a variety of underlying pathologies? Why assume that only internal variables rather than contextual variables can sufficiently explain patterns of observed consistency and similarity within the dispersion of symptoms and signs? Why are the symptoms and behaviors displayed by women suffering from this syndrome not seen as contextually disadvantaging rather than inherently disadvantaging? Why does society disapprove of these behaviors in women? Why are socio-cognitive factors related to the dominant gender ideologies not included in explanations of how women experience their bodies? Why is the body perceived as "a natural object," a relatively independent variable, rather than a dependent ideological variable? Who has been given the authority to interpret the female body and are those interpretations not influenced by misogynist assumptions and socio-cognitive factors that affect perception? Where is the place of female subjectivity (as a speaking subject) within the discourse of clinical language?

Although there are some feminist researchers[5] who are beginning to ask these questions, by and large these questions are very seldom voiced within PMS research. The reason for this is that such questions begin to challenge a fundamental perspective that organizes all disease-model theorizing, a perspective constituted by the medical gaze and its object of attention, "the medicalized body." This "medicalized body" can be characterized as follows:

> (a) It is perceived as an obstacle to the patient's will, either beyond the control of the patient or seriously interfering with desired psychological and behavioral normalcy, and as such further reified as an object that requires outside medical management.
>
> (b) Its malfunctioning is construed as clusters of symptoms and signs which are causally linked to discrete pathophysiological or psychogenic events.
>
> (c) It is perceived as a unit, as a relatively independent variable within ideological contexts, a mental/soma unit that can be manipulated and altered internally to regulate its normal functioning.

(d) It is perceived as a split unit, divided by the boundaries of psyche and soma, which are linked by linear causalities and mediated, however problematically, by neurosecretions of the brain and/or other hormonal chemistries.

From the perspective of a woman diagnosed with PMS and subjected to the scrutiny of the medical gaze, the "medicalized body" so constructed appears an alien object, subject to the authority of an "objective" observer who reads a story of pathological causality into her experience. As was pointed out earlier, the scaffolding of "symptoms" onto the subjective discourse of a woman's experience of her body or environment strips her words of their subjectively intended meaning. That she might be furious with someone, and justifiably so, is countenanced as a sign of "irritability" or "aggression." Likewise the medical construction of the body further strips the body from its social and ideological context, encasing it within the boundaries of the clinic. The body becomes problematized as an aggregate of clinically detectable organic or psychogenic events with causalities located within the boundaries of the psycho-soma unit. That the body might be symbolically mediated by social factors unnamed by clinical discourse is obscured once the body is stripped from its social and ideological contexts. Since both the clinician and the patient are not immune to the influence of these ideologies, it is not surprising that the "cleansing" function of medical discourse cannot rid itself of obscured ideological influences. Its imperfection is built into the social construction of its object of study: "the medicalized body," an object signified and fully enclosed by medical discourse.

An alternative view of the body perceives the body as symbolically mediated, as the site of many possible interpretations depending on the selective attentions and signifying practices of the observer, as an ideological variable, and as dependent on socio-cognitive factors which influence how the body and its internal states are interpreted. This is not to say that the body does not have inherent dysfunctions which cause pain, suffering, disability and death, nor that these dysfunctions may not be remedied by various modalities of intervention or healing. What I am claiming here is that the factors that have influence on how the body is interpreted may be of a much larger range than is countenanced by reductionist clinical discourse and by expanding that discourse a different construction of the body emerges. It becomes a site at least partially mediated by ideology and interpreted in conjunction with socio-cognitive factors that are internalized by the subject and those observing her behaviors.

A parallel can be seen between the way the female body is medicalized under PMS research and the body's placement within dominant ideological constructions. Within sexist ideology, the female body is often primarily defined by its utility and function within the reproductive and sexual strategies of patriarchal hegemony. In patriarchal cultures this results in an exploitation and appropriation of the female body and its powers, which in turn disempowers the female sex and devalues the necessary labor and energy she gives to others. Ideologically this act of appropriation is justified by the naming of difference: the sexes are perceived as inherently and significantly different in ways that account for the differences in the social distribution of power and exchange. These differences are not seen as relational, as embedded in the economy of human social relations of power and exchange, but as reified traits belonging to the body and derived from nature. The female body, so constructed and once stripped from the social nexus of relationships that confine and regulate its use, is perceived as a thing, which is further coded as defective and inferior to the male body. Thus female sex differences are reified as inherently disadvantaging. These attitudes internalized by subjects and observers are further reinforced by everyday observations of women's behavior, by expectations made about women, and even by women's self-negating behaviors and attitudes about themselves. The assumptions become metaphysical—always present and verifiable within descriptions of reality and irrefutable by counter examples which, if they exist, are given special exempt status or effectively discounted.

In PMS medical theory, similar reification occurs in positing causalities for female behavior within the boundaries of the body and in dislocating that body from the social contexts in which it is lived and interpreted. In many psychogenic and organic models of PMS etiology, there exist certain internal states within the menstrual cycle that are considered to be sufficient or jointly sufficient cause of the behavioral and physiological deviance observed in women. Given this model, if environmental or contextual factors are seen as influencing the biological variables of menstruation, they enter into consideration by route of psychogenic factors that either cause disruption in internal hormonal functioning or presuppose a malfunctioning pathophysiological background.

Other research studies related to self-attribution theory have challenged this reductionist conception of the body. In these studies, socio-cognitive factors, such as environmental cues, social expectations, and even language itself are considered significant if not overriding factors in the subject's interpretation of bodily states and in the self-attribution of certain symptoms. Two examples of such studies are summarized below:

> (1) Schachter and Singer (1962) gave injections of placebo and epinephrine (an arousal-producing substance) to subjects in three different groups: (a) one group was informed that the injection would cause arousal symptoms, such as sweating and rapid pulse (informed condition), (b) another group was told the injection would cause non-arousal conditions such as

itching and headache (misinformed condition) and (c) a third group was given no information about the effects of the injection (uninformed condition). The subjects were then placed in social contexts which elicited one of two emotional responses, anger or euphoria. The results showed that subjects in the informed group were unaffected by the emotional cues, since they explained the bodily changes as due to the injection. Uninformed and misinformed subjects experienced the emotion that was presented in environmental cues and did not attribute their feelings to the injection. Placebo-injected subjects did not experience any emotion, even in the presence of the emotional cues. Schachter and Singer concluded that an emotional state requires that the subject be in a state of physiological arousal and be given cues or emotional labels for interepreting the emotions. These cues enter into the subject's interpretation of the body as socio-cognitive factors.

(2) The influence of social cognitive factors is also indicated by Ruble's study (1977) of the reliability of self-report scales. In her study, 44 women undergraduates at Yale University were divided into three groups. The subjects were initially informed that they were part of a contraceptive study and that it was necessary to predict the day of their next menses using an electroencephalogram. After these tests, women in the first group were individually informed that they were in their premenstrum and would have their menses in one or two days. The women in the second group were individually informed that they were in the intermenstrual phase of their cycles and had at least a week to ten days before their next menses. The women in the third group were given no information regarding the test results. However, each woman actually received this information when she was in the sixth or seventh day before the onset of her next menses, a figure derived from each woman's menstrual history. This fact was unbeknownst to the women. After receiving the fake EEG interpretations, each woman was given a self-report Moos Menstrual Distress Questionnaire, with the result that women in the first group who believed they were in their premenstrum scored significantly higher than women in the second group. The third group (controls) scored between these two groups. This result suggests to Ruble that even self-report techniques, by far a somewhat reliable research tool, may give an overestimated measure of actual premenstrual distress, when social cues and expectations from the environment become intervening variables.

What these studies suggest is that such socio-cognitive factors may intervene on several levels, either in the hiatus between physiological states and the subject's interpretation of those states or in the subject's self-report observations of her physiological states.[6] These observations, which I have previously labeled Type A, are again filtered and codified into the discourse of symptoms and signs which gives measure to these initial observations in terms of deviance and implied morbidity. These derivative and more rarified observations (Type B) constitute an aggregate of signs and symptoms as components of the "medicalized body." Contrary to these reductionist models, Schachter and Singer's work suggests that socio-cognitive factors influence the emergence of primary data in subject attributions and that both physiological arousal states and contextual cues explain the existence of an emotionally labelled state, not physiological factors alone. Ruble's work[7] suggests that the contextual cues are so strong that subjects can experience expected emotional states associated with the premenstrum, even when the body is not in its premenstrual phase. These findings run contrary to medical theorizing based on "the medicalized body," which often views the menstrual cycle as an independent and sufficient cause of women's deviant behaviors. As is the case with ideological reification of the female body in everyday discourse, this tendency to locate etiologies within the body leads to further reification of a complex interactive process. Our interpretations of bodily states are not placed within symbolically mediated social relations, but occur within the displacement of clinical settings and reductionist medical discourse.

Breaking the grip of this discourse may create other possibilities. There is, for example, a more positive way of construing the strengths of female cyclicity. As Randi Koeske's ground-breaking work (1976, 1980, 1983) suggests, the premenstrual phase may open the female body to a state of nonspecific arousability rather than specific emotional states. The labelling of these states may be more dependent than we think on external environmental and cultural cues. Given my analysis, the labeling of these nonspecific states under male signifying practices is disadvantaging to females, precipitating female behaviors that seem to "fit" negative expectations and attitudes. I believe that it is better to view these nonspecific states as "windows of sensitivity," which grant the human female fine tuning with her environment. This gender-specific advantage of the female sex makes it possible for her to feel more deeply and more keenly during certain periods of the month, and if carefully developed and 'schooled,' such a biological advantage could be used as a power which adds more clarity and insight to human experience. However, this would require that the intentions and meanings of female voice not be swaddled by the masculinist discourse of morbidity and madness. It would also require a different cultural context or at least the demolition of the one we have.

Conclusion

Do my arguments suggest that we must return to the view that "it's all in your head?" This is not the case. The reductionism of medical research sets up a false dichotomy between explanations sufficiently grounded within the body vs. female fantasy. The experience of PMS is a reality for many women, and for some women it is extreme and severe. To be told that it's "all in your head" is both insulting and arrogant. These experiences are rooted in the body and its physiological mechanisms. However, to be told that it is due to a specific hormonal imbalance that can be medically managed is often overly simplistic and dangerously premature given the state of current research.

What I have argued in this essay is that we need to be at times leary of new PMS research and the assumptions behind its practice. The assumptions carry troublesome ideological content, when the codification of symptoms results in the morbidification of a sex difference which renders all women inherently disadvantaged in a man's world. In addition reductionist models further complicate this, by ignoring the impact of socio- cognitive variables. What needs to be acknowledged is that our experience of the body is very likely symbolically mediated by ideologies and socio-cognitive factors that impact on how one interprets bodily states.

An alternative view of the body suggests that the body can be seen as an ideological variable, a site of interpretation grounded in the corporeal but immersed in a wider sex/gender system (Rubin 1975; Harding 1983), which provides interpretive contexts for our experience of the body. To erase this context threatens to erase the meanings of subjective female voice; to displace woman's body into the singular domain of reductionist medical discourse; and to alienate woman from her body as an obstacle "out of control" and requiring medical management. So contructed, the "medicalized female body," a cyclically diseased entity, becomes *what woman is*. Furthermore, the body viewed as an object within a clinical setting, becomes the problem, the obstacle to be fixed. Thus woman is defined as requiring reparation. As in the case of hegemonic reification of "woman" as "body" in patriarchal ideology, this construction of "woman" adds legitimacy to the conservative notions of negative female essentialism.

To persist in marking a difference which commutes devaluation in the service of domination is to continue to build the scaffolds of that domination. It is no accident that a culture threatened by the erasure of gender differences as criteria for discrimination and domination would try to reinstate the social significance of such differences through its most legitimate authority—science. But it is also no accident that a culture based on domination would practice a science of domination and that the separation of this science from ideologies of masculinism and misogyny is indeed tenuous if not illusory. What I have suggested in this essay is that lines of influence between ideology and science need to be carefully scrutinized and the social use of science in the subjugation of women put through the lens of feminist criticism.

NOTES

1. Premenstrual Syndrome has been recently added to the appendix of the 1987 revised edition of the American Psychiatric Association's *Diagnostic and Statistical Manual III*. As an item in the appendix it lacks an official number usable for insurance purposes, but its presence as a descriptive category suggests such a likelihood in the future. This would further legitimate PMS as a clinical entity (Sandall 1987).

2. For purposes of this paper, I will replace the term "premenstrual syndrome" with "premenstrual changes" when referring to women's subjective reports and physiological changes associated with the premenstrum. I will retain the term "premenstrual syndrome" or "PMS" when discussing the research or findings that assume the existence of a syndrome and/or a disease-state substrate. My own terminology leaves open the question of how this periodicity and its aggravation are to be explained.

3. A notable exception is the research of Haskett (1980), among others.

4. For an overview of the research using organic models, see *Psychosomatics* (1985) 26 (10): 785-816; Abplanalp (1983); Reid and Yen (1981).

5. Feminist PMS researchers who have started to develop alternative explanations and experimental methods include Parlee, Aplanalp, Sherif, Koeske, among others.

6. The alternative socio-cognitive model suggested here differs from the psychogenic in a number of ways. The psychogenic model can be construed as one possible language for depicting emotional states. This language requires its own interrogation, since the environmental/contextual cues admitted by psychogenic explanations are heavily invested in psychoanalytic or psycho-therapeutic assumptions. Furthermore, in PMS psychogenic explanations, emotions placed within the context of personal psychological history carry causal significance for observed signs and symptoms. The focus is on this causality as well as on the meaning of the emotion or feeling. The socio-cognitive approach focuses more closely on the subject's description of phenomenologically experienced bodily states. By extension, the subject can be further understood as socially constructed in and through this process of interaction with language and other cultural heteronomies. For an analysis which begins to articulate this process from a radical materialist/feminist standpoint, see Frigga Haug (1987).

7. This interpretation of Ruble's results has been challenged by Ann Voda (1977) in an unpublished manuscript which bears comment. Voda's arguments include the following relevant points (a) the women subjects may have completed the questionnaires so as to validate findings of the EEG rather than describing their physiological and emotional states, (b) the young women, told that the research methodology had been successfully used on older women, may have experienced a subtle coercion to trust the pseudo-EEG results more than otherwise, (c) the time of ovulation was not checked by independent measures, so that some women in the first group could have been in their premenstrum. Voda's critique does raise intriguing questions regarding what was being described by the respondents in their self-reports, but her comments do not preclude the influence of socio-cognitive factors on women's self-assessment abilities. In fact, her critique indicates the further possibility that dependence on authoritarian discourse, expressed in medicine and scientific research, can profoundly affect women's self-assessment behaviors. This returns again to the question of how female subjectivity is socially constructed. (See note 6.)

REFERENCES

Abplanalp, J.M. 1983. Premenstrual syndrome: A selective review. *Women and health* 8:107-124.

Abplanalp, J.M., R.M. Rose, A.F. Donnelly and L. Livingson Vaughn. 1979. Psychoendocrinology of the menstrual cycle. II. The relationship between enjoyment activities, moods, and reproductive hormones. *Psychosomatic Medicine* 41 (8):605-615.

Allen, Hilary. 1984. At the mercy of her hormones: Premenstrual tension and the law. *m/f* 9:19-44.

Altmann, M., E. Knowles and H. Bull. 1941. A psychosomatic study of the sex cycle in women. *Psychosomatic Medicine* 3:199-225.

Arms, Suzanne. 1975. *Immaculate deception: A new look at women and childbirth in America.* Boston: Houghton Mifflin.

Boorse, C. 1975. On the distinction between disease and illness. *Philosophy and Public Affairs* 5 (1):49-68.

Brahams, Diana. 1981. Premenstrual syndrome: A disease of mind. *The Lancet* 2 (8257):1238-1240.

Coppen, A. and N. Kessel. 1963. Menstruation and personality. *British Journal of Psychiatry* 109:711—715.

Dalton, Katharina. 1960. Menstruation and accidents. *British Medical Journal* 2:1425-1426.

——— 1961. Menstruation and crime. *British Medical Journal* 2:1752-1753.

——— 1966. The influence of mother's menstruation on her child. *Proceedings of the Royal Society of Medicine* 59:1014.

——— 1977. *The premenstrual syndrome and progesterone therapy.* London: William Heinemann Medical Books.

——— 1979. *Once a month.* Great Britain: Hunter House, Inc.

Doering, C.H., H.C. Kraemer, H.K.H. Brodie and D.A. Hamburg. 1975. A cycle of plasma testosterone in the human male. *Journal of Clinical Endocrinology and Metabolism* 40:492-500.

Ettorre, Betsy. 1986. Women and drunken sociology. *Women's Studies International Forum* 9 (5):515-520.

Fortin, J.N. E.D. Wittkower and F. Katz. 1958. A psychosomatic approach to the premenstrual tension syndrome: A preliminary report. *Canadian Medical Association Journal* 79:978-981.

Glass, G.S., G.R. Heninger, M. Lansky and K. Talan. 1971. Psychiatric emergency related to the menstrual cycle. *American Journal of Psychiatry* 128:705-711.

Haire, Doris. 1974. *The cultural warping of childbirth.* New Jersey: International Childbirth Education Association.

Harding, Sandra. 1983. The visiblity of the sex/gender system. In *Discovering reality*, ed. Sandra Harding and Merrill Hintikka. Dordrecht, Holland: D. Reidel.

Harrison, Wilma, Judith Rabkin and Jean Endicott. 1985. Psychiatric evaluations of premenstrual change. *Psychosomatics* 26 (10):789-799.

Haskett, R.F., M. Steiner, J.N. Osmand, J.B. Carroll. 1980. Severe premenstrual tension: Delineation of the syndrome. *Biological Psychiatry* 15:121-139.

Haug, Frigga, ed. 1987. *Female sexualization* London: Verso.

Israel, S.L. 1938. Premenstrual tension. *Journal of the American Medical Association* 110:1721-1723.

Ivey, M.E. and J.M. Bardwick. 1968. Patterns of affective fluctuation in the menstrual cycle. *Psychosomatic Medicine* 30:336-345.

Janowsky, D.S., R. Gorney and B. Kelley. 1966. The curse: Vicissitudes and variations of the female fertility cycle. I. Psychiatric aspects. *Psychosomatics* 7:242-247.

Janowsky, D.S., R. Gorney, P. Castelnuovo-Tedesco, C.B. Stone. 1969. Premenstrual increases in psychiatric admission rates. *American Journal of Obstetrics and Gynecology* 103:189-191.

Kihlstroem, J.E. 1971. A male sexual cycle. In *Current problems in fertility*, ed. A. Ingelman-Sundberg and N.O. Lunell. New York: Plenum Press.

Koeske, Randi. 1976. Premenstrual emotionality: Is biology destiny? *Women and health* 1:11-14.

———1980. Theoretical perspectives on menstrual cycle research. In *The menstrual cycle: A synthesis of interdisciplinary research*, ed. A.J. Dan, E.A. Graham and C.P. Beecher. New York: Springer.

———1983. Lifting the curse of menstruation: Toward a feminist perspective on the menstrual cycle. *Women and health* 8:1-16.

Laws, S. 1983. The sexual politics of premenstrual tension. *Women's Studies International Forum* 6 (1):19-31.

Lieber, A. L. and C. R. Sherin. 1972. Homocides and the lunar cycle: Towards a theory of lunar influence on human emotional disturbance. *American Journal of Psychiatry* 129:69-74.

Luckhaus, Linda. 1985. A plea for PMT in the criminal law. In *Gender, sex, and the law*, ed. Susan Edwards. Great Britain: Croom Helm Ltd.

MacKinnon, P.C. and I.L. MacKinnon. 1956. Hazards of the menstrual cycle. *British Medical Journal* 1:555.

MacPherson, Kathleen. 1981. Menopause as disease: The social construction of a metaphor. *Advances in Nursing Science* 3:95-113.

Mandell, A. and M. Mandell. 1967. Suicide and the menstrual cycle. *Journal of the American Medical Association* 200:792-793.

McCance, R.A., M.C. Luff and E. E. Widdowson. 1937. Physical and emotional periodicity in women. *Journal of Hygiene* 37:571-611.

Moos, R.H. 1968. The development of the menstrual distress questionnaire. *Psychosomatic Medicine* 30:853-867.

——— 1969. A typology of menstrual cycle symptoms. *American Journal of Obstetrics and Gynecology* 103:390-402.

Moos, R.H., B.S. Kopell, F.T. Melges, I.D. Yalom, D.T. Lunde, R.B. Clayton and D.A. Hamburg. 1969. Fluctuations in symptoms and moods during the menstrual cycle. *Journal of Psychosomatic Research* 13:37-44.

Nicholson, John and Kay Barltrop. 1982. Do women go mad every month? *New Society* 59:226-228.

Parker, A.S. 1960. The premenstrual tension syndrome. *Medical Clinician North America* 44:339-341.

Parlee, Mary Brown. 1973. The premenstrual syndrome. *Psychological Bulletin* 80 (6):454-465.

——— 1974. Stereotypic beliefs about menstruation: A methodological note on the Moos menstrual distress questionnaire and some new data. *Psychosomatic Medicine* 36 (3):229-240.

——— 1982a. New findings: Menstrual cycles and behavior. MS 11 (3):126

——— 1982b. Changes in moods and activation levels during the menstrual cycle in experimentally naive subjects. *Psychology of Women Quarterly* 7 (2):119-131.

Paulson, M.J. 1961. Psychological concommitments of premenstrual tension. *American Journal of Obstetrics and Gynecology* 81:733-738.

Pennington, V.M. 1957. Premenstrual syndrome. *Journal of the American Medical Association* 164:638-643.

Premenstrual syndrome (special issue). 1985. *Psychosomatics* 26 (10):785-816.

Reid, Robert. and S.S. Yen, 1981. Premenstrual syndrome. *American Journal of Obstetrics and Gynecology* 139:84-104.

Ritenbaugh, Cheryl. 1982. Obesity as a culture-bound syndrome. *Culture, Medicine and Psychiatry* 6:347-361.

Rubin, Gayle. 1975. The traffic in women: Notes on the political economy of sex. In *Toward an anthropology of women*, ed. Rayna Reiter. New York: Monthly Review Press.

Ruble, D.N. 1977. Premenstrual syndrome: A reinterpretation. *Science* 197:291-292.

Sandall, Hilary. 1987. Feminism and psychiatry. Paper presented at the Centrum för Kvinnliga Forskare och Kvinnoforskning, Uppsala University, May 26.

Schacter, S. and J.E. Singer. 1962. Cognitive, social and physiological determinants of emotional states. *Psychology Review* 69:379 -399.

Sherif, C. W., L.A. Wilcoxon, S. L. Shrader. 1976. Daily self-reports on activities, life events, moods, and somatic changes during the menstrual cycle. *Psychosomatic Change* 38:399-417.

Seagull, E.A. 1974. An investigation of personality differences between women with high and low pre-menstrual tension. *Dissertation Abstracts International* 34 (9B):4675.

Sommer, Barbara. 1978. Stress and menstrual distress. *Journal of Human Stress* 4 (3):5-47.

—— 1982. Menstrual distress. In *The complete book of women's health*, ed. Gail Hongladarom, Ruther McCorkle and Nancy Woods. New Jersey: Prentice Hall.

Sondeheimer, S., E. Freeman, B. Scharlop and K. Rickels. 1985. Hormonal changes in premenstrual syndrome. *Psychosomatics* 26 (10):803-810.

Southam, A.L. and F.P. Gonzaga. 1965. Systemic changes during the menstrual cycle. *American Journal of Obstetrics and Gynecology* 91:142-165.

Stark, Evan, Ann Flitcraft, William Frazier. 1979. Medicine and patriarchal violence: The social construction of a private event. *International Journal of Health Services* 9 (3):461-93.

Suarez-Murias, E. 1953. The psychophysiological syndrome of premenstrual tension with an emphasis on the psychiatric aspect. *International Record of Medicine* 166:475-486.

Sutherland, H. and I.A. Stewart. 1965. A critical analysis of the premenstrual syndrome. *Lancet* 1:1180-1193.

Tonks, C.M., P.H. Rack and M.J. Rose. 1968. Attempted suicide and the menstrual cycle. *Journal Psychosomatic Research* 11:319-323.

Viet, H. 1955. Psychosomatic aspects of premenstrual tension. *Wisconsin Medical Journal* 54:599-601.

Voda, Ann. 1977. Unpublished manuscript.

Wickham, M. 1958. The effects of the menstrual cycle on test performance. *British Journal of Psychiatry* 49:34-41.

Whitehead, R.E. 1934. Women pilots. *Journal of Aviation Medicine* 5:47-49.

Women and the Mismeasure of Thought

JUDITH GENOVA

Recent attempts by the neurological and psychological communities to articulate thought differences between women and men continue to mismeasure thought, especially women's thought. To challenge the claims of hemispheric specialization and lateralization studies, I argue three points: 1) given more sophisticated biological models, brain researchers cannot assume that differences, should they exist, between women and men are purely a result of innate structures; 2) the distinction currently being drawn between verbal/spatial thinking abilities is fraught with ideological commitments that undermine the intelligibility of the distinction; 3) the model of thinking as information processing which underlies all this research confuses thinking with internal processing strategies.

Around the turn of the century, researchers were convinced that the five ounce difference in weight between male and female brains was the cause of female cognitive inferiority. Picturing the brain as a container and intelligence as some kind of vital fluid, they concluded that woman's smaller vessel could only hold less knowledge. Women were doomed to be less intelligent than men: their anatomy, wherever one looked, prevented it. Happily, in what soon came to be known as the "elephant problem," elephants and whales rescued women from this particular argument. If intelligence were a matter of absolute brain weight, elephants and whales would outscore men on intelligence tests handily. Since this was clearly absurd (species chauvinism remains unchanged today), absolute brain weight was quickly abandoned as a measure of intelligence. In its place various other relative measures were proposed: brain weight as an expression of body weight, body height, thigh bone weight or cranial height. The last alternative was promptly dismissed since it meant that Negroes, Australians and Eskimos would have the advantage; similarly, brain weight expressed as a function of body weight was disallowed since this gave women the advantage. The only alternative that gave men significantly heavier brains and thus preserved the already established prejudice was to express brain weight as a measure of body height.[1]

The above brief description of the history of craniometry has not been

Hypatia vol. 3, no. 1 (Spring 1988). © by Judith Genova.

exaggerated. The prejudice of the investigators and of the society that accepted them was as open as I have suggested; that is, the avowed reason for rejecting cranial height as a measure of brain weight was that it would make Eskimos smarter and according to the reigning ideology, that just could not be the case. While hindsight has made this prejudice visible to most current researchers, it has not discouraged them from seeking new and more ingenious measures of difference. Today, hemispheric specialization and laterality studies are dressing old arguments in new clothes. Given the nature of the differences they find between women and men and science's total domination of the authorial voice, the danger to women's lives has never been more threatening.

According to hemispheric specialization studies, women are considered to be left brain dominant, while men are right brain dominant: "Where such stylistic differences appeared, males tended to exhibit a style characteristic of right hemisphere dominance and females tended to exhibit left hemisphere processing" (Waber 1979, 173). The characterization of these "stylistic differences" is surprising given the traditional view of men's and women's rationality: "the left hemisphere is seen as verbal and analytic, processing items one at a time, while the right hemisphere is more holistic and spatial, processing items simultaneously" (Bryden 1979, 121). In another more detailed description, details versus wholes are emphasized along with the recognition that visuo-spatial stimuli are the domain at issue:

> The left hemisphere is oriented to detail. Acting on its own, it breaks a visual configuration into its component parts and attends to its internal features. The right hemisphere is oriented to the whole pattern of the configuration. (Waber 1979, 172)

One passage attempts to articulate these differences succinctly:

> In the search for a more fundamental mode of hemispheric specialization, several versions of an analytic (serial, focal, difference-detecting, time-dependent, sequential) versus holistic (parallel, diffuse, similarity-detecting, time-independent, spatial, global, synthetic or gestalt) dichotomy have independently surfaced. (Bradshaw 1980, 3:229)

What is so initially striking about these accounts is that the age-old tales of women being more intuitive and holistic, while men were thought to be more analytic and logical, are completely reversed. Somewhere, somehow the supposed special abilities of women and men have been switched. While I have not been able to determine exactly when this reversal, so typical of patriarchical history, has taken place, I suspect it is related to developments in computer technology. Perhaps it is no accident that men become intuitive and holistic just at the time when computers can successfully simulate logical, analytic thought. Another guess is that the right side which has always been identified with the male in most mythical and cultural accounts must remain superior, no matter what characteristics are associated with it (Needham 1973). Sacrificing logical and analytic thought is a small price to pay for remaining God's right hand man.

Naturally, now that men are thought to have intuitive qualities, they are honored. Spatial, holistic attributes have become associated with creative thinking, the kind of thinking needed for genius and especially scientific and mathematic creativity. Dr. William Blackmore, for example, was quoted as saying at the AAAS meeting:

> I simply don't believe that women have been locked out of brilliant creative accomplishments by social conditions. Male hormones, he suggested, result in superior right brain performance in men and it is this biological factor, not 'social conditions' that dictates the greater accomplishments of men in architecture, engineering, and art. (Quoted in Sayers 1982, 101)

Of course, it has not mattered and will not matter that women's presumed edge for analytic and relational skills has not meant their inclusion in all those fields requiring such abilities. Nor has it mattered that the percentage of women possessing holistic, spatial skills—the curves for these abilities overlap considerably—is not reflected in the numbers admitted into the scientific and mathematical professions today. Instead, women's ability in these areas is forgotten and men's spatial/holistic processing is crowned as the only path to true creativity. Once again men are viewed as natively equipped to do the truly inspired work. Women's cold, analytical, rational powers can only make them plodding amateurs in the creative game. Whatever the task, determinists will argue that men's holistic skills will assure them more efficient, more penetrating discoveries. The current attack on women, then, is confined to a kind of intelligence; it is specifically aimed at keeping them out of the world of science and trivializing their achievements in any field as routine and studied. It seems women are still missing five ounces, only now it is not the number of ounces that count, but the kind of ounces being discussed. At last, it has been granted that women have the stuff, just not the right stuff.[2]

The claims about lateralization, although they are part of the pattern, are not entirely synonymous with those of hemispheric dominance. Lateralization refers to the idea of greater or lesser specialization. Although the results in these studies are highly contested and contradictory, males are supposed to depend more on the left brain for language and conversely on the right brain for non-verbal skills; thus, they are thought to be more lateralized. Women, on the other hand, appear to be bilateral in that they do not show as much hemispheric specialization.[3] (The finding of a thicker corpus callosum in women presumably supports greater bilaterality, although there is no direct correlation between the anatomical data and the notion of laterality.) The issue here is, what is the significance of greater or less laterality? Suppose it were true that women were less lateralized? At most, it would be an advantage since they are less handicapped during certain kinds of brain lesions. Moreover, as Springer and Deutsch note, women's superior verbal skills should make less lateralization an asset:

> There is, of course, no logical reason to expect that greater lateralization necessarily leads to superior verbal ability. In fact, we have to assume the opposite to explain the superior verbal ability of females. According to behavioral tests and clinical data, women appear to be less lateralized for language functions, yet as a group they are superior to men in language skills. (1981, 129)

Nevertheless, there are those who claim that greater lateralization is better. Levy, for example, argues that specialization indicates a more "perfect evolution of the mechanisms involved" (Sayers 1982, 100).[4] Others have argued that greater lateralization is required for excellence in a particular area, for example, mathematics. In a section entitled, "Why Is There No Woman Beethoven?" Thomas R. Blakeslee, author of a popular book on split-brain theory, says:

> One of the disturbing paradoxes about the difference between men and women is that there have been (as yet) no women who are *towering* geniuses. . . . Certainly lack of opportunity or aspiration can account for much of this gap, but the towering genius is such a standout that poverty and persecution have seldom stood in his way. . . . Perhaps the answer to this paradox lies in woman's genetically faster rate of maturation and its resultant effect on brain organization. If a woman's brain tends to work like a pair of generalists while a man's is more like a pair of specialists, the mystery may be solved. (1983, 106-107)

In the past, men's greater variability was used to explain their superiority. For example, they are more likely than women to exemplify the extremes of intelligence; today, greater laterality is serving the exact same function. One frustrated researcher in the field concludes:

> We have seen that the evidence for sex-differential lateralization fails to convince on logical, methodological and empirical grounds. Is that surprising? Not all the points made in this critique are subtle, and some at least must be obvious to anyone in the field. Why then do reputable investigators persist in

> ignoring them? Because the study of sex differences is not like the rest of psychology. Under pressure from the gathering momentum of feminism, and perhaps in backlash to it, many investigators seem determined to discover that men and women 'really' are different. It seems that if sex differences do not exist, then they have to be invented. (Kinsbourne 1980, 3:242)

These claims raise a host of questions. Generally speaking, three broad areas are implicated by the questions: the nature of biological inquiry, the nature of difference, and the nature of thinking. To address some aspect of each of these areas, I shall focus on three specific questions. 1) If such differences in brain organization and function exist, to what extent can researchers assume that they represent innate, genetic biological differences? The tired nature/nurture controversy must be considered here because the very nature of brain studies provides a paradigm example of the inadequacy of this vexing dualism. 2) Do the differences, as they are currently being characterized by the scientific community, exist? Two different issues arise here: Is the distinction between relational/holistic or verbal/spatial real or ideological? If such a distinction were real, would it count as a *difference* in thought anyway? 3) What assumptions about the nature of thinking underwrite the whole project? Again, two related issues arise here. To what extent can thinking be measured as information processing or, more generally, is thinking a matter of internal brain activities, however they are construed? This last question does not repeat the issues of nature/nurture; rather, it hopes to override completely the discourse of science on the nature of thinking.

1) Given the very nature of brain investigations, most researchers assume they have discovered innate differences between the thinking of women and men. Conclusions are drawn from pathological cases and tests are given which measure functional performance. Interestingly, however, little anatomical data has been gathered to tie the detected differences to brain organization. There is no theory providing a mechanism for connecting sex differences in brain asymmetry and sex differences in cognitive behavior (Alper 1985, 20). Moreover, performance results have been obtained which contest the innateness assumption. For example, a study with black children from the rural south (Wittig 1979, 28) and one with Eskimos (Nash 1979, 282-283) did not show the same pattern of male/female competence. In an even more interesting study, Steven Vandenberg and Allan R. Ruse showed that "in general, greater sex-related differences on tasks with large spatial components were found in authoritarian cultures having rigidly defined sex roles" (1979, 71).

Along the same lines, both Margret Bryden and Diane McGuinness argue that "individual differences in performance on any measure of cerebral lateralization can be affected not only by differences in cerebral organization, but also by differences in attentional or information-processing strategies"

(Bryden 1980, 3:230; McGuinness 1980, 3:244). In other words, we tend to rely on modes of processing to which we have become accustomed and if women's early maturation of verbal skills predispose them to verbal processing, they may use it maladaptively for visuospatial tasks.[5] Rather than reflecting different structural organization, the tests may be only showing processing habits that can be changed by training. Thus, cultural and sociological factors determining sex roles can have an effect on hemispheric specialization patterns.

Thomas Bever's work on music provides yet another argument for the possible socialization of these patterns. In his attempt to argue that it is not skills per se which are lateralized, but styles of processing, he shows with respect to music, which was previously thought to be localized in the right brain, that the style of processing used is a function of one's training:

> This demonstration of the superiority of the right ear for music shows that it depends on the listener's musical experience; it demonstrates that the previously reported superiority of the left ear was due to the use of musically naive subjects, who treat simple melodies as unanalyzed wholes. (1980, 195)

While Bever himself does not think that his experiments prove that development is subject to experience—"It is also possible in principle that developing musical ability is not the cause of left-hemisphere dominance, but its result" (1980, 195)—his studies clearly undermine any simple faith that the way we view the world or the way we process it is simply a matter of unfolding genetic propensities. In addition, with respect to the issue of training, Vandenberg has found that practice with such tasks as model building increases the spatial scores of sixth grade girls. Thus, despite the easy assumption that brain studies are intimately linked to nature, there is reason to believe that learning factors play a large role in both hemispheric dominance and lateralization patterns.

These considerations are bolstered by the central argument against determinist positions, namely, that when a correlation is found between structures and functions, it can never be determined whether biology has produced it or whether the behavior has produced the biological correlate (Star 1979, 115). Thus, "there is no contradiction between the assertion that a trait is perfectly heritable and the assertion that it can be changed radically by environment" (Lewontin, Rose and Kamin 1984, 99). As some biologists stress, it is impossible to distinguish between the variables of nature and nurture since the process of development totally intertwines them. Indeed, Lewontin, Rose and Kamin will not even accept the new interactionist model of the dynamics of nature and nurture since such a model depends on there being two variables to interact. Instead, they argue that

> Organisms do not simply adapt to previously existing, autonomous environments; they create, destroy, modify, and internally transform aspects of the external world by their own life activities to make this environment. Just as there is no organism without an environment, so there is no environment without an organism.—Neither organism nor environment is a closed system; each is open to the other. (1984, 273)

For Steven Jay Gould, the real question is, Why would such a debate rage over the organ that is most labile, most flexible, and marks our species' departure from direct programmed response? It is ironic that determinists keep trying to fix what is most maleable (Gould 1984, 66-67). In a wonderful image, Lewontin, Rose and Kamin describe the frustrating resilience of determinist positions and inadvertently reveal the reason for it:

> Critics of biological determinism are like members of a fire brigade, constantly being called out in the middle of the night to put out the latest conflagration, always responding to immediate emergencies, but never with the leisure to draw up plans for a truly fireproof building. Now it is IQ and race, now criminal genes, now the biological inferiority of women, now the genetic fixity of human nature. All of these deterministic fires need to be doused with the cold water of reason before the entire intellectual neighborhood is in flames. (1984, 265)

While I cannot help but appreciate Lewontin, Rose and Kamin's faith in rationality, it is, given science's track record, less than realistic. They don't seem to realize that the firing synapses of the brain create an electrical fire which is inimical to reason's cold water. Moreover, with the advent of sociobiology the whole neighborhood is already in flames. The problem is that scientists disown their past by relegating it to "bad science" or to the poor scientific standards of the day, and fail to realize what I will call the banality of bad science. Throughout its history, especially in the human sciences, scientific investigation has been as subject to distortion and ideological infiltration as any other form of discourse. Instead of learning from the past, scientists continue to believe in their purity and thus perpetuate the same mistakes, especially in the area of sex and race differences. Clearly, fire-fighting techniques, other than unaided reason, must be employed. We must deconstruct the new dualism created by this research.

2) The second question I raised above was what are the ideological underpinnings of the verbal/spatial distinction? The first thing to note, as my initial quotations from hemispheric studies demonstrate, is that different researchers use different characterizations of the distinction. Springer and Deutsch

comment that while the verbal/visuo-spatial distinction is based on experimental evidence, all other designations seem more speculative (1981, 236). Yet, even this minimal characterization seems loaded. Maccoby and Jacklin were among the first to question the facile reduction of the minimal version of the distinction to verbal versus mathematical. They noted that one should expect a correlation between high verbal and mathematical skills since both require manipulation of symbols (1974, 63-65). And indeed, refined analysis has proved them right. According to the latest studies, most math skills are relational and occur in the left brain; only those math tasks involving non-analytic spatial visualization seem to be dominant in the right brain. For the purposes of my analysis, however, I will focus on Thomas Bever's version of the distinction since he attempts to refine the terms of the debate and offers sufficient information to make some assessments.

Objecting to the thought that the two hemispheres are best characterized in terms of skills such as verbal, spatial, he argues that what is really at issue is relational versus holistic processing. For him, language is left-hemisphered because it typically requires relational processing: "The left hemisphere is supposed to be specialized for propositional, relational and serial processing of incoming information, while the right hemisphere is more adapted for the perception of appositional, holistic and synthetic relations" (1980, 188). So, the question is, what does it mean for someone to be relational rather than holistic?

> For Bever, the difference is "intuitively clear": Holistic processing involves the direct association of a mental representation with a stimulus and response; relational processing involves at least two such associations, and the manipulation of a relation between the two mental representations. (1980, 189)

In other words, to be relational means to give an account or make a connection or comparison. And so in relational processing we are, as Bever suggests, comparing at least two representations. Holistic is much more problematic. Etymologically, it is a peculiar term since despite its spelling it is more related to "whole" (from OE. hal), complete, healthy, rather than "hole" (OE. hol), empty, hollow. Thus, like "holy" (OE. halig) it suggests a complete number of parts, divine, inviolate. To be holistic, then, is not only to see wholes, but to be whole. The implication of divine omniscience can be seen in Bever's use of the label. Like the Gestaltists, he uses it to name a process whereby one proceeds without going through steps; rather one has an immediate iconic grasp of a stimulus through the agency of one representation. There is no need to compare and contrast; one just sees. I can not help hearing Wittgenstein's discussions of sudden knowing or the "ah-ha" phenomenon in knowledge. Like Wittgenstein, I do not want to deny this experience; but I am not convinced that it involves a different form of processing. Wittgenstein, for example, argues that while it seems to come from out of the blue, it has nevertheless been prepared for in the inquiry process. There may be a leap in the end, but familiar discursive processes have prepared for its discovery.

One more passage illustrates the mystification of Bever's position, and of the others who depend on this distinction:

> The notes of a chord are 'related' to each other; indeed, the chord depends for its character on such a relation. But perceiving a chord is not a relational act in the technical sense of the term. While the perception involves a relation between two notes, it does not require separate identifications of the two notes independently *and* in relation to each other. . . . For example, why is recognizing a rotated or displaced figure a holistic task; such recognition presupposes a relation just as a chord presupposes separate notes (physical movement). The answer is that the object is set only in relation to its (identical) self; therefore the rotation task does not meet the 'found' criteria of relational processing. (1980, 203)

The claim is that holistic processing is not discursive, but depends on something like an unmediated recognition. Intuition whether it is assigned to women or men remains incorrigible. The representation is taken as the identical mirror of the object. Thus, holistic processing escapes the mediation of culture and gives a genuine, undistorted picture of the universe. Finally, the hope of epistemology from its very beginning is finding its scientific legitimation. Yet anyone who participates in these mental rotation tests recognizes that they are comparing representations and contrasting possible alternatives to obtain the correct answer even if they are not experiencing Bever's robot-like description of relational processing.[6] Besides, the very terms used to name the dichotomy, "whole" versus "part," along with its reliance on a superstitious dichotomy which sees the right as good and bright and the left as sinister and dull, reveal its ideological intentions. Consequently, I doubt that so-called spatial processing is a different mode of processing. Even Bever says that it presupposes relations; rather, I suspect it is the same one that certain people, because of habituation, can do faster than others. And as has always been the case, it is speed of response, rather than quality of response that measures success on intelligence tests. Just as parallel processing is only a faster version of serial processing in computers, holistic processing is a faster version of relational processing in humans. At the very least we can argue, according to Bever's claims, that successful holistic processing depends on the perceiving of relations.

Another way of addressing the issue of difference is to ask the question:

Would different modes of processing count as a basic difference in thought between women and men? I think not. As Sybil Wolfram argues: "The answer 'no' will then be appropriate where the terms are used in a fashion so varied that they cannot be said to mark a single division, where they mark one of degree and not of kind, as with 'rich' and 'poor', 'large' and 'small' or where there is not a single principle of differentiation as in the case of 'incest' and 'adultery'" (1973, 359). Essentially, the claim of a basic difference of thought needs the presence in one group of a trait which is absent in the other. As M.J. Morgan argues, "Sexual dimorphism refers to a state of affairs in which individual members of the two sexes can be distinguished by one or more features, such as the secondary sexual characteristics. On the other hand, characteristics such as height could never be used for determining whether an individual were male or female" (1980, 244). Similarly, verbal and spatial processing could never individuate a population. Everyone has both processes and the degree of overlap between women and men is far too great to substantiate a claim of two different modes of thought. The difference is one of degree and thus according to the above criteria does not justify a claim to being a basic difference in thought.

3) The last question concerns the relationship between information processing and thinking. While the latest philosophical positions argue that thinking is nothing but information processing, the issue is anything but settled. The importance of this debate for the issues of this paper should be clear: those of us who question current cognitive science claims that thinking is a matter of information processing will not find the discovery of two different modes of processing, should they exist, to be very significant.

To begin a discussion of this question, consider the following example: Two researchers from Stanford have done some work on mapping styles (i.e., how men and women remember and map their familiar terrains). The results tended to follow the proposed differences in processing strategies: "Men had a greater topographic sense of the campus terrain, placing buildings more accurately with respect to spatial coordinates. On the other hand women demonstrated a more accurate sense of distance. . . . Females focus more on the landmarks and the distance between individual elements, and males focus more on the topographic network of roads and other connectors which provide a geometric framework for the location of buildings" (McGuinness and Sparks 1983, 96-99). Thus the females negotiated the details of the campus; they knew what the buildings were and had a more accurate sense of their distance from each other. Men knew their relative positions and had a sense of the lay of the land. Thus, they knew different things in the sense that they had different information. Men, while they had a generalized orientation, had no idea which building was which or how long it would take them to get from one building to another. If they were hungry, only trial and error could get them to the cafeteria. Women on the other hand would be lost in giving specific directions for getting from point A to point B. So the two different ways of framing space have consequences for our understanding of something as familiar as our local habitat.

Yet, how consequential is this? The processing strategies in affecting what we know may lead to different thoughts. However, what makes us identify thinking with the processing strategies? Perhaps the processing makes thinking possible; however, it seems more reasonable to view the thinking as a second order reflection on the information obtained rather than as the process producing the information. All sorts of processes are programmed into the computer to produce information, but few would say that the computer thinks. Thinking is something done with the information in context, not by the information. Having different information may yield different ideas; it may affect what we think. But to claim that these different processing strategies are themselves different forms of thinking is to confuse the process with the product. It is as if we said that seeing is identical with the physiological processes that make seeing possible. What we see and how we see in the sense of a perspective or frame of reference cannot be measured by the information structuring devices of the eye and brain. However, the issue here is larger than the identification of thinking with information processing. The question that must be asked is, Can thinking be measured by any sort of physiological or psychological process?

References to how people think have always been subject to a systematic ambiguity. Sometimes, as Clifford Geertz recognizes, we are speaking about thinking as a verb or process, other times, as a noun or product (1983, 147). Confusion between the two senses is significant since they are radically different. Thinking processes are thought of as internal mental mechanisms such as structural properties of the brain or embedded symbolic codes. They are perceived of as internal, private, biologically innate and causal with respect to the way we think or the thoughts we have. The second sense of "think," namely, as product (i.e., thoughts, ideas or systems of beliefs) refers to a way or mode of thinking, or as Geertz says, a culture. Unlike the processes of thinking, thoughts are external, public, socially sensitive and acausal with respect to the processes of thinking.

The first sense of "think," that of process or act, has been dominant in the West. Not only is it the default interpretation, but only variations in thinking processes have counted as official measures of difference. As a rule, while phenomenal differences are the ones that lead people to suspect thought differences in the first place, they have been discounted as measures of difference. In the past, if the sense of difference could not be confirmed by finding empirical differences in processes, one would either have to abandon the claim of difference or retreat to hidden, mysterious differences that will be detected someday. The reason for these alternatives is that the usual way of understanding the relationship between the two senses of "think" is to see the processes of thinking as the cause of the thoughts. Given this assumption, it becomes especially difficult to argue that, for example, women and men differ in their way of thinking, but not in their processes of thinking.

Yet, this is precisely what anthropologists, who have long speculated on thought differences, have been forced to do with respect to race and cultural differences: "In particular, we want to emphasize our major conclusion that *cultural differences in cognition reside more in the situations to which particular processes are applied than in the existence of a process in one cultural group and its absence in another*" (Cole *et al.* 1971, 233). Instead of abandoning the claim of difference, anthropologists are maintaining it while at the same time denying differences in thought processes. The same situation, I believe, is occurring in feminism. Feminists, like myself, are claiming differences in the way women and men think, but most of us do not mean, nor do we believe, that these differences can be traced finally to differences in brain processes.[7] However, as Geertz worries, this dual response threatens a coherent discussion of the issue of difference:

> The overall movement of those sciences [social] during that period [from the twenties till now] has been one in which the steady progress of a radically unific view of human thought, considered in our first, 'psychological' sense as internal happening, has been matched by the no less steady progress of a radically pluralistic view of it in our second, 'cultural' sense as social fact. And this has raised issues that have now so deepened as to threaten coherence. (1983, 148)

The threat to coherence that Geertz is referring to is reflected in the following reasoning: if thoughts are conceived as the products of thinking and if thinking is conceived as the product of internal brain processes, then we cannot both acknowledge thought differences on the second level and deny internal differences on the first level. Such a strategy severs the connection between thinking and thoughts.

For Geertz, something has got to give and for him, this something is our conception of thinking as an internal psychological process. Calling for an "ethnography of thought," Geertz wants to develop an "outdoor psychology" which believes that "ideation, subtle or otherwise, is a cultural artifact" (1983, 153). As he says, "It is a matter of conceiving of cognition, emotion, motivation, perception, imagination, memory . . . whatever, as themselves, and directly, social affairs" (1983, 153). While Wittgenstein urged such a transformation of our concept of thinking years ago, perhaps the time is right to hear it.[8]

Consider any other human activity, from walking to talking. In no case would one argue that the styles of our walks or talks, no less the content of the latter, are caused by the physiological processes underlying them. True, these processes are conditions of our activities in that they make certain options possible, but they are not the causes of them. Geertz is right; what has to give is our conception of thinking as an inner psychological process. Thinking must come out of the closet of the mind and enter the social and cultural world as both its product and producer.

If the presumed causal relation between thinking as process and thinking as thought can be severed—and I believe it can and must be—it becomes possible to understand Carol Gilligan's work, for example, as identifying differences in thought without being forced to produce their biological correlates. As Gilligan notes in her conversation with children and adults, women's sense of self as well as their interactions with others is built around the need to remain connected and avoid hurting others. Like the mythic natives of tribal cultures, Gilligan's female respondents have trouble answering questions about hypothetical situations because they do not accept general rules of fairness that negotiate such situations for Western men. For them, "the world coheres through human connection rather than through systems of rules" (1982, 29). Thus, everything turns on the situation and the particular constellation of factors that comprise it. Antigone simply cannot live in Creon's world; she cannot live by his rules, not only because of their tyrannical nature, but because they are insensitive to the situation at hand. She will bend the rules to sustain life, or in this case, bury the dead; mercy overrides justice. The claims of equality which demand that all people be seen as potential values for a variable in any judgment, of fairness which demand an even exchange and those of justice which ask that the rules be upheld at any cost have been the trademark of male Western society; yet they are not the first court of appeal for women. Of course, women learn these rules, just as men in later life come to learn the practices of caring. Yet, their different approaches do mark a difference in thinking.

I particularly like this example because there are no bad guys in it. The ethic of justice, fairness and rights is not something we can dismiss as evil. And Gilligan recognizes that the two ethics must temper each other. Moreover, Gilligan is careful not to tie her argument to any kind of biological determinism. She says:

> The different voice I describe is characterized not by gender but theme. Its association with women is an empirical observation, and it is primarily through women's voices that I trace its development.But its association is not absolute, and the contrasts between male and female voices are presented here to highlight a distinction between two modes of thought and to focus a problem of interpretation rather than to represent a generalization about either sex. (1982, 2)

That women speak in this voice is a contingent matter, not a necessary one;

it has been their experience as the primary caretakers of children, not their genes, that has given rise to this voice. Thus, nothing prevents the sexes from learning each other's ways. However, Gilligan's main point is that there is something other than the moral analysis of and by men which has been enshrined in the texts of the West. Moreover, the defect in women's moral personality as it was diagnosed by Freud, Kohler and Piaget, was more a defect in their own moral perceptions than in the women they were doctoring.

Women, because of their experiences in the world, are different from men. The "corporeal ground" of their intelligence is not their bodies per se, as Adrienne Rich believes, but the body of culture they inhabit. To recognize these thought differences and to properly credit their social origins is to begin the task of producing an outdoor psychology. Thinking, in the sense of beliefs and belief systems, must be severed from the internal mechanisms of the brain, if we are ever to understand and respect the differences among people. Above all, we must resist the latest attempt to reduce differences to internal thinking mechanisms.

NOTES

I am grateful to The Center for the Humanities at Wesleyan University for the opportunity to research this project during my appointment as Senior Research Associate in the fall of 1984.

1. For more detailed critiques of this history see Lewontin, Rose and Kamin (1984); Janet Sayers (1982); Steven Jay Gould (1981); ElizabethFee (1980, 415-433).

2. For further criticisms of these studies see Ann Fausto-Sterling (1985); Joseph Alper (1985); Ruth Bleier (1986).

3. There are two competing models of lateralization: The Buffery/Gray hypothesis (1972, 123-157) argues in contrast to the Levy/Sperry (1972, 159-180) hypothesis described in the text that women are more lateralized rather than bilateral.

4. Surprisingly, Levy is a woman. In reading this literature, I have been struck by the number of women involved in this research. In my short bibliography alone there are 10. Since some of these women are among the first to draw sexist conclusions, someone must work on the Phyllis Schlafly phenomenon in science research.

5. Deborah Waber argues that right and left brain specializations are directly correlated with late and early maturation. The later we mature, the more the right brain will be specialized for spatial tasks. Since women, on average, mature earlier than men, there will be a sex difference in this area (Waber 1979, 179). Vandenberg (1979) has criticized her results because of the poor correlation between physical and mental growth.

6. The tests used to measure spatial ability involve mentally rotating or orienting objects in space. They depend on the ability to manipulate objects or images in three-dimensional space without appeal to verbal or other unmotivated signs.

7. See Carol Gilligan (1982); Sandra Harding and Merill B. Hintikka (1983); Adrienne Rich (1976); Mary Daly (1982); Elizabeth Abel (1982); Alison M. Jaggar (1983, 371-377); Elaine Marks and Isabelle de Courtivron (1980).

8. For contemporary arguments of this position see Jeff Coulter (1979, 1983).

REFERENCES

Abel, Elizabeth, ed. 1982. *Writing and sexual difference*. Chicago: University of Chicago Press.

Alper, Joseph. 1985. Brain asymmetries. *Feminist Studies* 11 (1): 7-37.

Bever, Thomas. 1980. Broca and Lashley were right: Cerebral dominance is an accident of growth. In *Biological studies of mental processes*, ed. David Kaplan. Massachusetts: M.I.T. Press.

Blakeslee, Thomas R. 1983. *The right brain*. New York: Berkley Books.

Bleier, Ruth. 1986. Sex differences research: Science or belief? In *Feminist approaches to science*, ed. Ruth Bleier. New York: Pergamon Press.

Bradshaw, John L. 1980. Sex and side: A double dichotomy interacts. *The Behavioral and Brain Sciences* 3:229-230.

Bryden, M.P. 1979. Evidence for sex-related differences in cerebral organization. In *Sex-related differences in cognitive functioning: Developmental issues*, ed. Michele Andrisen Wittig and Anne C. Petersen. New York: Academic Press.

———. 1980. Sex differences in brain organization: Different brains or different strategies. *Behavioral and Brain Sciences* 3:230-231.

Buffery, Anthony and Jeffery A. Gray. 1972. Sex differences in the development of spatial and linguistic skills. In *Gender differences: Their ontogeny and significance*, ed. Christopher Ounsted and David D. Taylor. Edinburgh: Churchill & Livingstone.

Cole, Michael, John Gay, Joseph A. Glick and Donald Sharp. 1971. *The cultural context of learning and thinking: An exploration in experimental anthropology*. New York: Basic Books.

Coulter, Jeff. 1979. *The social construction of mind*. New Jersey: Roman and Littlefield.

———. 1983. *Cognitive theory revisited*. New York: St. Martin's Press.

Daly, Mary. 1982. *Gyn/ecology*. Boston: Beacon Press.

Fausto-Sterling, Anne. 1985. *Myths of gender*. New York: Basic Books.

Fee, Elizabeth. 1980. Nineteenth century crainology: The study of the female skull. *Bulletin of the History of Medicine* 53:415-433.

Geertz, Clifford. 1983. *Local knowledge*. New York: Basic Books.

Gilligan, Carol. 1982. *In a different voice*. Cambridge, MA: Harvard University Press.

Gould, Steven Jay. 1981. *The mismeasure of man*. New York: W.W. Norton.

———. 1984. Biological potentiality vs. biological determinism. In *Designer genes: I.Q. ideology and biology*, ed. Chee Heng Leng and Chan Chee Khoon. Maylasia: Institute for Social Analysis.

Harding, Sandra and Merill B. Hintikka. 1983. *Discovering reality*. Dordrecht, Holland: D. Reidel.

Hubbard, Ruth. 1979. Have only men evolved? In *Women look at biology looking at women*, ed. Ruth Hubbard, Mary Sue Henifin and Barbara Fried. Cambridge, MA: Schenkkman.

Jaggar, Alison M. 1983. *Feminist politics and human nature*. New Jersey: Rowman and Allanheld.

Kinsbourne, Marcel. 1980. If sex differences in brain lateralization exist, they have yet to be discovered. *Behavioral and Brain Sciences* 3:241-242.

Levy, Jerre. 1972. Lateral specialization of the human brain: Behavioral manifestations and possible evolutionary basis. In *The biology of behavior*, ed. J.A. Kiger, Jr. Corvallis: Oregon State University Press.

Lewontin, R.C. and Steven Rose and Leon Kamin. 1984. *Not in our genes*. New York: Pantheon Books.

Maccoby, Eleanor Emmons and Carol Nagy Jacklin. 1974. *The psychology of sex differences*. California: Stanford University Press.

McGuinnes, Diane. 1980. Strategies, demands, and lateralized sex differences. *Behavioral and Brain Sciences* 3:244.

McGuinness, Diane and Janet Sparks. 1983. Cognitive style and cognitive maps in representations of a familiar terrain. *Journal of Mental Imagery* 7 (2):91-100.

Marks, Elaine and Isabelle de Courtivron. 1980. *New French feminisms*. Amherst: University of Massachusetts Press.

Morgan, M.J. 1980. Influences of sex on variation in human brain asymmetry. *Behavioral and Brain Sciences* 3:244-245.

Nash, Sharon Churnin. 1979. Sex-role as a mediator of intellectual functioning. *In Sex-related differences*. See Bryden 1979.

Needham, Rodney, ed. 1973. *Right and left: Essays on dual symbolic classification*. Chicago: University of Chicago Press.

Rich, Adrienne. 1976. *Of woman born*. New York: W.W. Norton.

Sayers, Janet. 1982. *Biological politics: Feminist and anti-feminist perspectives*. New York: Tavistock.

Springer, Sally P. and Georg Deutsch. 1981. *Left brain, right brain*. New York: W.H. Freeman.

Star, Susan Leigh. 1979. Sex differences and the dichotomization of the brain: Methods, limits, and problems in research on consciousness. In *Genes and gender II: Pitfalls in research on sex and gender*, ed. Ruth Hubbard and Marian Lowe. New York: Gordian Press.

Vandenberg, Steven and Allan R. Ruse. 1979. Spatial ability: A critical review of the sex-related major gene hypothesis. In *Sex-related differences*. See Bryden 1979.

Waber, Deborah. 1979. Cognitive abilities and sex-related variations in the maturation of the cerebral cortical functions. In *Sex related differences*. See Bryden 1979.

Wittig, Michele Andrisen. 1979. Genetic influences on sex-related differences in intellectual performance: Theoretical and methodological issues. In *Sex-related differences*. See Bryden 1979.

Wolfram, Sybil. 1973. Basic differences of thought. In *Modes of thought*, ed. Robin Horton and Ruth Finnegan. London: Faber and Faber.

Bibliography

This bibliography is designed to serve as a resource for research on feminism and science. Although not exhaustive, it contains representative works from three areas of investigation: works about the lives and status of women scientists; critiques of gender bias in the sciences; and feminist perspectives on the epistemology and metatheory of science.

Women in Science

Aldrich, Michele. 1978. Women in science. *Signs* 4 (1):126–135.

*Alic, Margaret. 1986. *Hypatia's heritage: A history of women in science from antiquity through the nineteenth century*. Boston: Beacon Press.

American Association for the Advancement of Science. 1976. *The double bind: The price of being a minority woman in science*. Washington, D.C.

Ames, Elinor. 1981. The status of women in Canadian psychology: A study of women in science. *International Journal of Women's Studies* 4 (4):431–40.

Arnold, Lois Barber. 1984. *Four lives in science: Women's education in the nineteenth century*. New York: Schocken.

Baldwin, R. S. 1981. *The fungus fighters: Two women scientists and their discovery*. New York: Cornell University Press.

Blackstone, Tessa and Helen Weinreich-Haste. 1980. Why are there so few women scientists and engineers? *New Society* 51:385–85.

Bluemel, Elinor. 1959. *Florence Sabin: Colorado woman of the century*. Boulder: University of Colorado Press.

*Briscoe, Anne. 1981. Diary of a mad feminist chemist. *International Journal of Women's Studies* 4 (4):420–30.

Briscoe, Anne M. and Sheila M. Pfafflin, eds. 1979. *Expanding the role of women in the sciences*. New York: The New York Academy of Sciences.

*Brooks, Paul. 1972. *The house of life: Rachel Carson at work*. Boston: Houghton Mifflin.

Bunting, Mary. 1965. *Women and the science professions*. Cambridge, MA: MIT Press.

Chepelinsky, Ana Berta, et. al. 1980. Women in chemistry. In *Science and Liberation*, ed. Rita Arditti et al, 257–266. Boston: South End Press.

An earlier version of this bibliography was published in 1985 in *Frontier* 8 (3):73–78. My thanks to Barbara Imber for her help with this bibliography.

*For those readers desiring an initial introduction to the issues of feminism and science, I recommend the starred items.

Chinn, P. Z. 1980. *Women in science and mathematics: Bibliography*. Arcata, CA: Humboldt State University.

Cole, Jonathan. 1979. *Fair science: Women in the scientific community*. New York: Free Press.

Curran, Libby. 1980. Science education: Did she drop out or was she pushed? In *Alice through the microscope: The power of science over women's lives*, ed. Brighton Women and Science Group, 22–41. London: Virago Press.

Frieze, Irene Hanson. 1978. Psychological barriers for women in sciences: Internal and external. In *Covert discrimination and women in the sciences*, ed. Judith Ramaley, 65–95. Boulder, CO: Westview Press.

Gleasner, Diana C. 1984. *Breakthrough: Women in science*. New York: Walker and Company.

Goodfield, June. 1981. *An imagined world: A story of scientific discovery*. New York: Harper & Row.

*Gornick, Vivian. 1983. *Women in science: Portraits from a world in transition*. New York: Simon and Schuster.

Haas, Violet B. and Carolyn C. Perrucci, eds. 1984. *Women in scientific and engineering professions*. Ann Arbor: The University of Michigan Press.

Haber, Lois. 1979. *Women pioneers of science*. New York: Harcourt, Brace, Jovanovich.

Hamilton, Alice. 1972. *Exploring the dangerous trades: The autobiography of Alice Hamilton*. Boston: Houghton Mifflin.

*Harrison, Michelle, M. D. 1983. *A woman in residence*. New York: Penguin Books.

Hinton, Kate. 1976. *Women and science: Science in a social context*. Manchester, England: Manchester University Press.

Hoobler, Icie Gertrude Macy. 1982. *Boundless horizons: Portrait of a pioneer woman scientist*. Pompano Beach, FL: Exposition Press.

Humphreys, S., ed. 1982. *Women and minorities in science: Strategies for increasing participation*. Boulder, CO: Westview.

Kahle, Jane Butler. 1982. *Double dilemma: Minorities and women in science education*. West Lafayette, IN: Purdue University.

Keller, Evelyn Fox. 1977. The anomaly of a woman in physics. In *Working it out: 23 women writers, artists, scientists, and scholars talk about their lives and work*, ed. Sara Ruddick and Pamela Daniels. New York: Pantheon Books.

*———. 1983. *A feeling for the organism: The life and times of Barbara McClintock*. San Francisco: W. H. Freeman.

Kistiakowsky, Vera. 1980. Women in physics: Unnecessary, injurious, and out of place. *Physics Today* 33 (2):32–40.

Kohlstedt, Sally Gregory. 1978. In from the periphery: American women in science, 1830–1880. *Signs* 4 (1):81–96.

Land, Barbara. 1981. *The new explorers: Women in Antarctica*. New York: Dodd and Mead.

Lonsdale, Kathleen. 1970. Women in science: Reminiscences and reflections. *Impact of Science on Society*, 20 (1):45–59.

Lovejoy, Esther Pohl, M. D. N.d. *Women physicians and surgeons: National and international organizations*. New York: Livingston Press.

*Malcom, Shirley Mahaley, Paula Quick Hall and Janet Welsh Brown. 1975. *The double bind: The price of being a minority woman in science*. Washington, D. C.: American Association for the Advancement of Science.

Martin, Ben and John Irvine. 1982. Women in science: The astronomical brain drain. *Women's Studies International Forum* 5 (1):41–68.

Mary Ingraham Bunting Institute of Radcliffe College. 1980. *Choices for science: Proceedings of symposium*. Cambridge, MA: M. I. Bunting Institute.

Mead, Margaret. 1972. *Blackberry winter: My earlier years*. New York: William Morrow.

Menninger, Sally Ann and Clare Rose. 1980. Women scientists and engineers in American academia. *Signs* 3 (3):292–99.

Meuron-Landolt, Moniquede. 1975. How a woman scientist deals professionally with men. *Impact of Science on Society* 25 (2):147–52.

Moore, Emily C. 1980. *Woman and health: United States 1980*. Washington, D.C.: U.S. Government Printing Office.

Morantz-Sanchez and Regina Markell. 1985. *Sympathy and science: Women physicians in American medicine*. New York: Oxford University Press.

Murphy, Angela Corigliano. 1980. Ladies in the lab. In *Science and liberation*, ed. Rita Arditti et al, 247–256. Boston: South End Press.

Noble, Lois. 1979. *Contemporary women scientists of America*. New York: Messner.

Patterson, Elizabeth Chambers. 1983. *Mary Somerville and the cultivation of science, 1815–1840*. Boston: Nijhoff.

*Ramaley, Judith A., ed. 1978. *Covert discrimination and women in the sciences*. Boulder, CO: Westview Press.

Richer, D., ed. 1982. *Women scientists: The road to liberation*. London: Macmillian.

Roark, Anne. 1980. Women in science: Unequal pay, unsold ideas, and sometimes unhappy marriages. *Chronicle of Higher Education* 20:3–4.

Rossi, Alice S. 1965. Women in science: Why so few? *Science* 148 (3674):1196–201.

*Rossiter, Margaret. 1982. *Women scientists in America: Struggles and strategies to 1940*. Baltimore: Johns Hopkins University Press.

Rossiter, Margaret. 1978. Sexual segregation in the sciences: Some data and a model. *Signs* 4 (1):146–51.

Ruddick, Sara and Pamela Daniels, eds. 1977. *Working it out: 23 women writers, artists, scientists, and scholars talk about their lives and work*. New York: Pantheon Books.

*Sayre, Anne. 1975. *Rosalind Franklin and DNA*. New York: W. W. Norton.

Schiebinger, Londa. 1987. The history and philosophy of women in science: A review essay. *Signs* 12 (2):305–32.

Scott, Joan Pinner. 1981. Science subject choice and achievement of females in Canadian high schools. *International Journal of Women's Studies* 4 (4):348–61.

Shapley, Deborah. 1975. Obstacles to women in science. *Impact of Science on Society* 25 (2):115–23.

Sheinin, Rose. 1981. The rearing of women for science, engineering, technology. *International Journal of Women's Studies* 4 (4):339–47.

Siegel, P. J. and K. T. Finley. 1985. *Women in the scientific search: An American bio-biography, 1724–1979*. Metuchen, NJ: Scarecrow Press.

Smith, Elske. 1978. The individual and the institution. In *Covert discrimination and women in the sciences*, ed. Judith Ramaley, 7–35. Boulder, CO: Westview Press.

Special Issue. 1978. Women, science, and society. *Signs: Journal of Women in Culture and Society* 4 (1).

Spieler, Carolyn, ed. 1977. *Women in medicine—1976: Report of a Macy conference*. New York: Josiah Macy, Jr. Foundation.

Stark-Adamec, Cannie. 1981. Practical tips for coping with the problems of being a 17 career person. *International Journal of Women's Studies* 4 (4):441–56.

Tereshkova-Nikolayeva, Valentina. 1970. Women in space. *Impact of Science on Society* 20 (1):5–12.

Vetter, Betty. 1976. Women in the natural sciences. *Signs* 1 (3, Part 1):713–20.

Walsh, Mary Roth. 1977. Doctors wanted: No women need apply. *Sexual barriers in the medical profession*, 1835–1975. New Haven: Yale University Press.

Weisstein, Naomi. 1979. Adventures of a woman in science. In *Women looking at biology looking at women*, ed. Ruth Hubbard et al, 187–203. Cambridge, MA: Schenkman.

White, Martha. 1970. Psychological and social barriers to women in science. *Science* 170 (3956):413–16.

Yee, C. Z. 1977. Do women in science and technology need the women's movement? *Frontiers* 2:125–28.

Yost, Edna. 1984. *Women of modern science*. Westport, CT: Greenwood Press.

Feminist Critiques of Sexism in the Practice of Science

Ann Arbor Science for the People Editorial Collective. 1977. *Biology as a social weapon*. Ann Arbor, Michigan.

Barker-Benfield, G. S. 1978. *Horrors of the half-known life: Male attitudes toward women and sexuality in 19th century America*. New York: Harper Colophon.

Bart, Pauline. 1977. Biological determinism and sexism: Is it all in the ovaries? In *Biology as a social weapon*, ed. Ann Arbor Science for the People Editorial Collective, 69–83. Ann Arbor, Michigan.

Birke, Lynda. 1980. From zero to infinity: Scientific views of lesbians. In *Alice through the microscope*, Brighton Women and Science Group, 108–123. London: Virago Press.

Birke, Lynda and Sandy Best. 1980. The tyrannical womb: Menstruation and menopause. In *Alice through the microscope*, Brighton Women and Science Group, 89–107. London: Virago Press.

Birke, L. and J. Silverton, eds. 1984. *More than the parts: Biology and politics*. London: Pluto Press.

Black, Datha Clapper. 1979. Displaced: The midwife by the male physician. In *Women looking at biology looking at women*, ed. Ruth Hubbard et al, 83–101. Cambridge, MA: Schenkman.

Bleier, Ruth. 1976. Myths of the biological inferiority of women: An exploration of the sociology of biological research. *University of Michigan Papers in Women's Studies*, 2 (2):39–63.

Bleier, Ruth. 1986. Lab coat: Robe of innocence or klansman's sheet? In *Feminist Studies/Critical Studies*, ed. Teresa de Lauretis, 55–66. Bloomington: Indiana University Press.

*Boston Women's Health Book Collective. 1984. *The new our bodies, ourselves: A book by and for women*. New York: Simon and Schuster.

Brighton Women and Science Group. 1980. *Alice through the microscope: The power of science over women's lives*. London: Virago.

Cole, Jonathan. 1980. Meritocracy and marginality: Women in science today and tomorrow. In *Choices for science*, 4–26. Cambridge, MA: M. I. Bunting Institute.

Conrad, Peter, Rochelle Kern and Joseph W. Schneider. 1980. *Deviance and medicalization: From badness to sickness*. St. Louis: C. V. Mosby Co.

*Conway, Jill. 1970. Stereotypes of femininity in a theory of evolution. *Victorian Studies* 14 (1):47–62.

Corea, Gena. 1977. *The hidden malpractice: How American medicine treats women as patients and professionals*. New York: William Morrow.

Couture-Cherki, Monique. 1980. Women in physics. In *Ideology of/in the natural sciences*, ed. Hilary Rose and Steven Rose, 206–216. Boston: G. K. Hall.

Culpepper, Emily. 1979. Exploring menstrual attitudes. In *Women looking at biology looking at women*, ed. Ruth Hubbard et al, 135–160. Cambridge, MA: Schenkman.

Dagg, A. I. 1983. *Harem and other horrors: Sexual bias in behavioural biology*. Waterloo, Ontario: Otter Press.

Dreifus, Claudia, ed. 1977. *Seizing our bodies: The politics of women's health.* New York: Vintage Books.

D'Onofrio-Flores, Pamela and Sheila Pfafflin. 1982. *Scientific-technological change and the role of women in development.* Boulder, CO: Westview Press.

Druss, Vicki and Mary Sue Henifin. 1979. Why are so many anorexics women? In *Women looking at biology looking at women*, ed. Ruth Hubbard et al, 126–133. Cambridge, MA: Schenkman.

*Ehrenreich, Barbara and Deirdre English. 1972. *Witches, midwives, and nurses: A history of women healers.* Old Westbury, NY: The Feminist Press.

———. 1974. *Complaints and disorders: The sexual politics of sickness.* Old Westbury, NY: The Feminist Press.

*———. 1978. *For her own good: 150 years of the experts' advice to women.* New York: Doubleday.

Falkner, Wendy. 1980. The obsessive orgasm: Science, sex and female sexuality. In *Alice through the microscope*, Brighton Women and Science Group, 139–162. London: Virago Press.

Fausto-Sterling, A. 1985. *Myths of gender: Biological theories about women and men.* New York: Basic Books.

Fee, Elizabeth. 1973. The sexual politics of Victorian social anthropology. *Feminist Studies* 1 (3–4):23–39.

———. 1976. Science and the woman problem: Historical perspectives. In *Sex differences: Social and biological perspectives*, ed. Michael Tietelbaum, 175–223. New York: Doubleday.

*———. 1980. Nineteenth century craniology: The study of the female skull. *Bulletin of the History of Medicine* 53:415–33.

———. 1983. *Women and health: The politics of sex in medicine.* New York: Baywood Publishing Co.

Feldman, Jacqueline. 1975. The savant and the midwife. *Impact of Science on Society* 25 (2):125–35.

Fisher, Elizabeth. 1979. *Woman's creation: Sexual evolution and the shaping of society.* New York: McGraw-Hill.

Fried, Barbara. 1979. Boys will be boys will be boys: The language of sex and gender. In *Women looking at biology looking at women*, ed. Ruth Hubbard et al, 37–59. Cambridge, MA: Schenkman.

Gilligan, Carol. 1982. *In a different voice: Psychological theory and women's development.* Cambridge, MA: Harvard University Press.

Goddard, N. and M. S. Henifin. 1984. A feminist approach to the biology of women. *Women's Studies Quarterly* 12:11–18.

Gould, Stephen Jay. 1978. Women's brains. *New Scientist* 80 (1127):364–66.

———. 1981. *The mismeasure of man*. New York: W. W. Norton.

Lambert, Helen H. Biology and equality: A perspective on sex differences. *Signs* 4(1):97–117.

Leacock, Eleanor. 1983. Ideologies of male dominance as divide and rule politics: An anthropologist's view. In *Woman's nature: Rationalizations of inequality*, ed. Marian Lowe and Ruth Hubbard, 111–121. New York: Pergamon Press.

Leiboweitz, Lila. 1978. *Females, males, and families: A biosocial approach*. London: Duxbury Press.

———. 1983. Origins of the sexual division of labor. In *Women's nature: Rationalizations of inequality*, ed. Marian Lowe and Ruth Hubbard, 123–147. New York: Pergamon Press.

*Lowe, Marian. 1983. The dialectic of biology and culture. In *Woman's nature: Rationalizations of inequality*, ed. Marian Lowe and Ruth Hubbard, 39–62. New York: Pergamon Press.

Lowe, Marian and Ruth Hubbard, ed. 1983. *Women's nature: Rationalizations of inequality*. New York: Pergamon Press.

Maccoby, Eleanor. 1970. Feminine intellect and the demands of science. *Impact of Science on Society* 20 (1):13.

Magner, Lois. 1978. Women and the scientific idiom: Textual episodes from Wollstonecraft, Fuller, Gilman, and Firestone. *Signs* 4 (1):61–80.

Martin, M. Kay and Barbara Voorhies. 1975. *Female of the species*. New York: Columbia University Press.

Merchant, Carolyn. 1980. *The death of nature: Women, ecology, and the scientific revolution*. San Francisco: Harper & Row.

Messing, Karen. 1983. The scientific mystique: Can a white lab coat guarantee purity in the search for knowledge about the nature of women? In *Women's nature: Rationalizations of inequality*, ed. Marian Lowe and Ruth Hubbard, 75–88. New York: Pergamon Press.

Morgan, Elaine. 1972. *The descent of woman*. New York: Stein and Day.

Newman, L. M., ed. 1984. *Men's ideas/women's realities: Popular science, 1870–1915*. Elmsford, NY: Perrgamon.

Quinn, Naomi. 1977. Anthropological studies on women's status. *Annual Review of Anthropology* 6:181–225.

Reed, Evelyn. 1978. *Sexism and science*. New York: Pathfinder Press.

Reid, Inez Smith. 1975. Science, politics, and race. *Signs* 1 (2):397–422.

Rogers, L. J. 1983. Hormonal theories for sex differences: Politics disguised as science. *Sex Roles* 9:1109–14.

Ruse, Michael. 1981. *Is science sexist? And other problems in the biomedical sciences*. Dordrecht: D. Reidel.

Ruzek, Sheryl Burt. 1978. *The women's health movement: Feminist alternatives to medical control*. New York: Praeger.

Sayers, Janet. 1980. Psychological sex differences. In *Alice through the microscope*, Brighton Women and Science Group, 42–61. London: Virago Press.

*———. 1982. *Biological politics: Feminist and anti-feminist perspectives*. New York: Tavistock Publications.

Scully, Diana. 1980. *Men who control women's health: The mis-education of obstetrician-gynecologists*. Boston: Houghton Mifflin.

Seaman, Barbara and Gideon Seaman, M. D. 1977. *Women and the crisis in sex hormones*. New York: Rawson Associates Publishers, Inc.

Shields, Stephanie. 1978. Sex and the biased science. *New Scientist* 80 (1132):752–54.

———. 1982. The variability hypothesis: The history of a biological model of sex differences in intelligence. *Signs* 7 (4):769–97.

Spanier, Bonnie. 1980. Critical filters in science careers. In *Choices for science* 33–41. Cambridge, MA: M. I. Bunting Institute.

Star, Susan Leigh. 1979. The politics of right and left: Sex differences in hemispheric brain asymmetry. In *Women looking at biology looking at women*, ed. Ruth Hubbard et al, 61–74. Cambridge, MA: Schenkman.

Tobach, Ethel and Betty Rosoff. 1978. *Genes and gender 1: On hereditarianism and women*. New York: Gordian Press.

Walberg, Herbert J. 1969. Physics, femininity, and creativity. *Developmental Psychology* 1 (1):47–54.

Walsh, Mary Roth. 1979. The quirls of a woman's brain. In *Women looking at biology looking at women*, ed. Ruth Hubbard et al, 103–125. Cambridge, MA: Schenkman.

Wertz, Dorothy and Richard Wertz. 1977. *Lying-in: A history of childbirth in America*. New York: Free Press.

Science and Feminist Epistemology

Addelson, Kathryn Pyne. 1983. The man of professional wisdom. In *Discovering reality*, ed. Sandra Harding and Merrill Hintikka, 165–186. Dordrecht, Holland: D. Reidel.

Arditti, Rita. 1980. Feminism and science. In *Science and liberation*, ed. Rita Arditti et al, 350–368. Boston: MA: South End Press.

Arditti, Rita, Pat Brenman and Steve Cavrak, eds. 1980. *Science and liberation*. Boston, MA: South End Press.

*Bleier, Ruth 1984. *Science and gender: A critique of biology and its theories on women*. New York: Pergamon Press.

*———, ed. 1986. *Feminist approaches to science*. New York: Pergamon Press.

Bonner, Jill. 1980. The cult of objectivity in the physical sciences. In *Choices for science*, 52–57. Cambridge, MA: M. I. Bunting Institute.

Bordo, Susan. 1986. The cartesian masculinization of thought. *Signs* 11 (3):439–56.

Easlea, Brian. 1981. *Science and sexual oppression: Patriarchy's confrontation with woman and nature*. London: Weidenfeld and Nicolson.

———. 1983. *Fathering the unthinkable: Masculinity, scientists and the nuclear arms race*. London: Pluto Press.

Fausto-Sterling, Anne. 1981. Women and science. *Women's Studies International Quarterly* 4 (4):41–50.

Fee, Elizabeth. 1981a. Is feminism a threat to objectivity? *International Journal of Women's Studies* 4 (4):378–92.

———. 1981b. Is there a feminist science? *Science and Naure* 4:46–57.

———. 1983. Women's nature and scientific objectivity. In *Women's nature: Rationalizations of inequality*, ed. Marian Lowe and Ruth Hubbard, 9–27. New York: Pergamon Press.

Fowlkes, Diane L. and Charlotte S. McClure, eds. 1984. *Feminist visions toward a transformation of the liberal arts curriculum*. Alabama: The University of Alabama Press.

Goodman, Madeleine J. and Lenn Evan Goodman. 1981. Is there a feminist biology? *International Journal of Women's Studies* 4 (4):393–413.

*Griffin, Susan. 1978. *Woman and nature: The roaring inside her*. New York: Harper & Row.

Haraway, Donna. 1978. Animal sociology and a natural economy of the body politic. *Signs* 4 (1):21–60.

*———. 1981. In the beginning was the word: The genesis of biological theory. *Signs* 6 (3):469–81.

*Harding, Sandra. 1981. The norms of social inquiry and masculine experience. *PSA 1980*, ed. P. D. Asguith and R. N. Giere, 305–324. Ann Arbor: Edwards Bros.

———. 1982. Is gender a variable in conceptions of rationality? *Dialectica* 36 (2–3):225–42.

*———. 1986. *The science question in feminism*. Ithaca: Cornell University Press.

———. 1986. The instability of the analytic categories of feminist theory. *Signs* 11 (4):645–64.

Harding, Sandra and Jean O'Barr, eds. 1987. *Sex and scientific inquiry*. Chicago: University of Chicago Press.

*Harding, Sandra and Merrill B. Hintikka, eds. 1983. *Discovering reality: Feminist perspectives on epistemology, metaphysics, methodology, and philosophy of science*. Dordrecht, Holland: D. Reidel.

Hartsock, Nancy. 1981. Social life and social science: The significance of the naturalist intentionalist dispute. In *PSA 1980*, ed. P. D. Asquith and R. N. Giere, 325–345. Ann Arbor: Edwards Bros.

Hein, Hilde. 1981. Women and science: Fitting men to think about nature. *International Journal of Women's Studies* 4 (4):369–77.

Irigaray, Luce. 1985. *Speculum of the other woman.* New York: Cornell University Press.

Keller, Evelyn Fox. 1980a. Feminist critique of science: A forward or backward move? *Fundamenta Scientiae* 1:341–49.

———. 1980b. Baconian science: A hermaphroditic birth. *The Philosophical Forum* 11 (3):299–308.

———. 1981. Women and science: Two cultures or one. *International Journal of Women's Studies* 4 (4):362–68.

———. 1982. Feminism and science. *Signs* 7 (3):589–602.

———. 1983a. Feminism as an analytic tool for the study of science. *Academe* 69 (5):15–21.

———. 1983b. Gender and science. In *Discovering reality*, ed. Sandra Harding and Merrill Hintikka, 187–205. Dordrecht, Holland: D. Reidel.

*———. 1985. *Reflections on gender and science.* New York: Yale University Press.

———. 1986. Making gender visible in the pursuit of nature's secrets. In *Feminist Studies/Critical Studies*, ed. Teresa de Lauretis, 67–77. Bloomington: Indiana University Press.

Koertge, Noretta. 1981. Methodology, ideology and feminist critiques of science. In *PSA 1980*, ed. P. D. Asquith and R. N. Giere, 346–359. Ann Arbor: Edwards Bros.

Longino, Helen. 1981 (Fall). Scientific objectivity and feminist theorizing. *Liberal Education* 67.

———. 1983a. Beyond bad science: Skeptical reflections on the value-freedom of scientific inquiry. *Science, Technology and Human Values* 8:7–17.

———. 1983b (March). Scientific objectivity and the logics of science. *Inquiry.*

*Longino, Helen and Ruth Doell. 1983. Body, bias, and behavior: A comparative analysis of reasoning in two areas of biological science. *Signs* 9 (2):206–27.

Lowe, Marian. 1981. Cooperation and competition in science. *International Journal of Women's Studies* 4 (4):362–68.

Mouton, Janice. 1983. A paradigm of philosophy: The adversary method. In *Discovering reality*, ed. Sandra Harding and Merrill Hintikka, 149–164. Dordrecht, Holland: D. Reidel.

Overfield, K. 1981. Dirty fingers, grime, and slag heaps: Purity and the scientific ethic. In *Men's studies modified*, ed. Dale Spender, 237–48. Elmsford, NY: Pergamon.

Purdy, Laura M. 1986. Nature and nurture: A false dichotomy. *Hypatia: A Journal of Feminist Philosophy* 1 (1):167–174.

*Rose, Hilary. 1983. Hand, brain, and heart: A feminist epistemology for the natural sciences. *Signs* 9 (1):73–90.

*Rosser, Sue. 1982. Androgyny and sociobiology. *International Journal of Women's Studies* 5 (5):435–444.

———. 1984. A call for feminist science. *International Journal of Women's Studies* 7 (1):3–9.

———. 1985. Introductory biology: Approaches to feminist transformations in course content and teaching practice. *Journal of Thought* 20 (3):205–17.

———. 1986. *Teaching science and health from a feminist perspective*. New York: Pergamon Press.

Stehelin, Liliane. 1980. Sciences, women and ideology. In *Ideology of/in the natural sciences*, ed. Hilary Rose and Steven Rose, 217–230. Boston: G. K. Hall.

Tosi, Lucia. 1975. Women's scientific creativity. *Impact of Science on Society* 25 (2):105–14.

Tuana, Nancy. 1983. Re-fusing nature/nurture. *Hypatia* 3, published as a Special Issue of *Women's Studies International Forum* 6 (6):621–32.

———. 1985. Re-presenting the world: Feminism and the natural sciences. *Frontiers* 8 (3):73–78.

———. 1986. A Response to Purdy. *Hypaia: A Journal of Feminist Philosophy* 1 (1): 175–178.

Wallsgrove, Ruth. 1980. The masculine face of science. In *Alice through the microscope*, Brighton Women and Science Group, 228–240. London: Virago.

Whitbeck, Caroline. 1984. A different reality: Feminist ontology. In *Beyond domination*, ed. Carol Gould, 64–88. New York: Rowman and Allanheld.

Woodhull, Ann, Nancy Lowry and Mary Sue Henifin. 1985. Teaching for change: Feminism and the sciences. *Journal of Thought* 20 (3):162–73.

Notes on Contributors

LINDA ALCOFF received her Ph.D. in philosophy from Brown University and is assistant professor of philosophy at Syracuse University, where she teaches continental philosophy and feminism. She is the mother of two children. She is currently working on a book on coherence and truth.

THE BIOLOGY AND GENDER STUDY GROUP was established in 1985 as a credited course in the biology department of Swarthmore College. This essay is a position paper developed in discussions with the original members of this group. *Athena Beldecos* and *Nancy Niemczyk* are presently seniors at Swarthmore College. *Nancy Niemczyk* has a program in biology and religion and is active in the Nuclear War Education Project as well as in campaigns protesting our involvement in Nicaragua. *Athena Beldecos* is pursuing a program in biology and English literature. *Lori Kenschaft* is presently executive director of the Association for Women in Mathematics. She majored with distinction in both biology and religion, as well as coordinating the AIDS educational program, a forum on women's concerns and the folk dance organization. She plans on entering the Unitarian Universalist ministry. *Andrew Wedel* is presently a graduate student in public health at the University of California, San Francisco. *Stephanie Schaertel* is pursuing graduate studies in chemistry at Cornell University. *Rebecca Rosenberg*, the instigator of this group, is presently studying dance in Boston. *Sarah Bailey* is a research associate in Bethesda, Maryland. *Karen Hicks* is a research technician at MIT. *Scott Gilbert* is an associate professor of biology who received his Ph.D. in biology and his MA in the history of science from Johns Hopkins University. He is the author of the textbook *Developmental Biology*. His research concerns molecular and cellular aspects of gene regulation, as well as the history and sociology of biological science.

CAROL MASTRANGELO BOVÉ teaches at Westminister College, New Wilmington, PA. She has written several articles on twentieth-century literary criticism and fiction. Her current project is a book on Julia Kristeva for the Twayne Series.

JUDITH GENOVA is a professor of philosophy and feminism at Colorado College. She has published articles on Wittgenstein and on various issues in aesthetics. She recently edited a book on feminist theory, *Power, Gender, Values* (1987), published by Academic Printing and Publishing, Edmonton, Canada.

RUTH GINZBERG is a visiting assistant professor at Indiana University at South Bend. Her doctoral work in philosophy was at the University of Minnesota, where she was in the first class of graduate students to receive graduate level minors through the Center for Advanced Feminist Studies. Her recent work has been on developing a theory and pedagogy of feminist logic. She has a teenage daughter who was born at home, in the gynocentric tradition.

SANDRA HARDING is professor of philosophy and director of women's studies at the University of Delaware. Her most recent books are *The Science Question in Feminism* (Cornell, 1986), and two edited collections: *Sex and Scientific Inquiry*, with Jean O'Barr (Chicago, 1987), and *Feminism and Methodology: Social Science Issues* (Indiana, 1987).

LISA HELDKE completed her Ph.D. in 1986 at Northwestern University. Her dissertation, from which her article is taken, is entitled *Coresponsible Inquiry: Objectivity from Dewey to Feminist Epistemology*. She is currently a visiting assistant professor at Gustavus Adolphus College in St. Peter, Minnesota.

RUTH HUBBARD is a professor of biology at Harvard University, where she teaches courses dealing with the interactions of science and society, particularly as they affect women. She has written numerous articles for professional and general audiences and has edited several books. She is a member of Science for the People, the National Women's Health Network, and the Committee for Responsible Genetics.

LUCE IRIGARAY holds a doctorate in linguistics and in philosophy. She is a practicing psychoanalyst who divides her professional time between writing, her private practice as an analyst, and research at the Centre Nationale de la Recherche Scientifique devoted to the relation of language and psychology. Her latest books translated into English are *This Sex Which is Not One* (Cornell, 1985) and *Speculum of the Other Woman* (Cornell, 1985).

EVELYN FOX KELLER has worked in theoretical physics, molecular and mathematical biology. She is currently professor of rhetoric and women's studies at the University of California at Berkeley, and is perhaps best known for her two books, *A Feeling for the Organism: The Life and Work of Barbara McClintock* (W. H. Freeman, 1983), and *Reflections on Gender and Science* (Yale University Press, 1985). Her most recent work has been on language and ideology in evolutionary theory.

HELEN E. LONGINO is an associate professor of philosophy at Mills College. She is the author of many articles in the philosophy of science and in feminist philosophy. She is co-editor of *Competition: A Feminist Taboo?* (The Feminist Press, 1987). Her book on social values and scientific inquiry is forthcoming from Princeton University Press in 1989.

ELIZABETH POTTER is associate professor of philosophy at Hamilton College. She has published articles in mainstream epistemology but has recently turned her attention to feminist approaches to epistemology and the philosophy of science; in this connection, she is exploring the gender politics in seventeenth-century science. She has been Executive Secretary of the Society for Women in Philosophy, Eastern Division, and is one of the founders of the Kirkland Historical Studies of Science and Technology conference series.

SUE V. ROSSER is the director of women's studies at the University of South Carolina at Columbia. She also holds an appointment as associate professor of preventive medicine and community health in the USC Medical School. Formerly she was chair of the division of theoretical and natural sciences and coordinator of women's studies at Mary Baldwin College. She received her Ph.D. degree in zoology in 1973 from the University of Wisconsin—Madison and while a post-doctoral fellow began teaching in the women's studies program at the University of Wisconsin during the first year of its existence in 1976. Since then she has taught courses in both biology and women's studies programs at the University of Wisconsin—Madison and Mary Baldwin College. She has authored several publications dealing with the theoretical and applied problems of women and science and authored the books *Teaching About Science and Health from a Feminist Perspective: A Practical Guide* and *Resistances of the Science and Health Care Professions to Feminism*, both published by Pergamon Press. As a consultant for the Wellesley Center for Research on Women, she has worked with faculty at several institutions that are attempting to include the new scholarship on women in the science curriculum.

NANCY TUANA is an associate professor in the history of ideas at the University of Texas at Dallas. She has published in the areas of philosophy of science and feminist theory, and is currently completing a book entitled *The Misbegotten Man: Scientific, Religious, and Philosophical Images of Woman's Nature*. She is editor of the APA *Newsletter on Feminism and Philosophy*.

JACQUELYN N. ZITA is a feminist educator and philosopher working as assistant professor of Women's Studies, University of Minnesota. She has recently completed an exchange professorship at the Centrum för Kvinnliga Forskare och Kvinnoforskning at Uppsala University in Sweden. Her current work includes editorship of *Matrices* and a series of essays on feminist epistemology, theories of sex/gender embodiment, and lesbian issues.

Index

Absolutism: coresponsible option, 104
Accountability: feminist science, 54; science and society, 121
Achillini, Alessandro: anatomy and reproductive theory, 159–60
Activity: masculinity and cell biology, 181; gender and metaphor in organic chemistry, 183. *See also* Creativity
Aeschylus: sexism and reproductive theory, 151, 167
Albertus Magnus: anatomy and reproductive theory, 159
Alchemy: Boyle and chemistry, 132
American Association for the Advancement of Science: status of women in science, 5
Anabolism: models of sex determination, 173, 174
Anatomy: Galen and reproductive theory, 154; primacy of male generative powers, 158–61, 163; brain size, 211–12
Androcentrism: science and objectivity, 8; sex hormones and behavior, 51–52; traditional Western science, 69, 71, 74; language and science, 81–82; sexism, 145n
Anglici, Ricardi: anatomy and reproductive theory, 159
Animalculism: preformation doctrine, 164, 165, 166–68
Anthropology: feminist theories, 10; male bias, 125; research and context, 127; cultural differences and thought processes, 222
Aristotle: gender politics, 140; sexist bias and reproductive theory, 147–53, 158, 159, 168, 172, 173, 174, 176; animalculist theory, 167
Art: gynocentric science, 71
Assumptions: methodology, 22; feminist inquiry, 28–29; values and hypotheses, 49; sex hormones and behavior, 51; network model, 143–44; gender bias, 172; ideology and premenstrual syndrome research, 205
Astruc, Jean: preformation doctrine, 168
Atomic theory: Anne Conway, 144. *See also* Corpuscular theory
Authority: science and women, 40–41, 42
Avicenna: anatomy and dissection, 161

Battery: premenstrual syndrome and observational bias, 191
Behavior: sociobiology and status of women, 6; sex hormones, 51–52, 53; science and society, 120; deviance and premenstrual syndrome, 188–89, 192, 197, 202; sexist bias and premenstrual syndrome research, 190–91
Benedetti, Allesandro: anatomy and reproductive theory, 160
Bias: science and gender, x–xi; scientific method, 23; values and science, 50, 51–52, 53–54; social science, 88; feminism and theory-choice, 89; sociobiology, 125; Aristotle and reproductive theory, 147–53; anatomy, 161, 163; gender and biology, 169; cellular and molecular biology, 179; feminist criticism, 183, 184; premenstrual syndrome and observational language, 190–97; premenstrual syndrome and explanatory models, 198–204
Biological determinism: status of women, 6–7; feminist approach, 8; environmental influences, 216–17
Biology: methods and androcentric assumptions, 17, 18; feminist method, 19–20, 30; feminist epistemologies, 23–24; gender, 27; politics and research, 27; subject and non-neutrality, 63; sex differences and discrimination, 122–23; male bias, 125, 169; sexism and reproductive theory, 172; feminist critique, 173, 179, 183–84; textbooks and gender-related images, 175, 185n; sexualization of the cell, 180; determinism, 216–17
Biotechnology: politics and science, 140–41
Birth defects: Aristotle and reproductive theory, 152–53
Blacks: family and feminist inquiry, 28
Boccaccio, Giovanni: history of science, 4
Body: ideology and premenstrual syndrome research, 201–202, 204, 205
Boerhaave, Hermann: anatomy and sexist bias, 163; animalculist theory, 167
Boston Women's Health Book Collective: women and science, 129
Botany: gynocentric science, 72
Boyle, Robert: theories and values, 49; scientific tradition, 113; story model and biography, 132–33; natural magic tradition and Comenius, 135; gender politics and

Boyle—*continued*
philosophy of science, 136–37, 139, 140; assumptions and network model, 143–44
Brain: Aristotle and sexist bias, 148; size and intelligence, 11–12. *See also* Craniometry; Hemispheric specialization; Intelligence; Lateralization

Campanella: natural magic tradition and political dissent, 134–35
Carson, Rachel: holistic approach, 70–71
Cesarean section: mortality and hospitals, 78–79
Charleton, Walter: gender politics and philosophy of science, 139–40
Chemistry: history of science, 9; Boyle and alchemy, 132; fertilization metaphors, 182–83
Childbirth: androcentric approach, 75–76; health and mortality, 77; hospitals, 77–79; midwifery and obstetrics, 79–81
Christianity: primacy of male generative powers, 158
Class: science and society, 120, 123
Coherence: theory-choice, 91, 92; conditions and assumptions, 143, 145n
Comenius, Jan: natural magic tradition and Boyle, 135
Communal property: gender politics and philosophy of science, 134
Communication: inquiry, 111, 113
Computers: hemispheric specialization studies and sexism, 212; thinking processes, 221
Conformity: institutions and feminist science, 55–56
Consciousness-raising: feminist method, 18–19
Context: objectivity and scientific method, 126–27
Contraception: science and language, 60
Conway, Anne: atomic theory, 144
Cooke, J.: animalculist theory, 167
Cooking: gynocentric science, 71
Coresponsible option: epistemological tradition, 104–105; correspondence theory of truth, 114n
Corpuscular theory: gender politics and philosophy of science, 135–40, 144
Correspondence: truth and theory-choice, 95, 101n; theory of truth, 114n-115n
Courtship: metaphor and sexism in reproductive theory, 174, 179
Craniometry: science and status of women, 6; history and prejudice, 211–12. *See also* Brain; Intelligence
Creativity: sexism and reproductive theory, 151–52, 168; science, 183; sexism and hemispheric specialization studies, 213. *See also* Activity
Criticism, feminist: politics and science, viii; recent scholarship, 6–8; methods and gender, 17–18; assumptions and gender bias, 172–73; cellular and molecular biology, 179; creativity and natural science, 183–84
Culture: science, xi, 120; gender and power, 42–43; knowledge and choice, 54; difference and thought processes, 222, 224
Curie, Marie: history of science, 74
Curriculum: feminists and science, 4; history of science, 5
Cyclicity: physiological systems, 189; observational bias and menstruation, 192, 197, 204; explanatory models of premenstrual syndrome, 198, 199
Cytoplasm: sexualization of the cell, 179–82

Darwin, Charles: feminist criticism, 6
Darwin, Erasmus: preformation doctrine, 167–68
Determinism. *See* Biological determinism
Dewey, John: epistemological tradition, 104; ontology, 105–108, 113–14; dualism, 109; inquiry, 110–11, 112
Difference: women as scientists, 35–36; Barbara McClintock, 37, 108, 110; duality and universality, 39–40, 41, 42; gender, 43; inquiry, 112; Aristotle's biology, 148; patriarchy and female body, 202; nature of thinking, 215; thinking processes, 221–22
Discourse: theory-choice, 92, 93, 94–95; premenstrual syndrome research, 200, 201, 204
Discrimination: women as scientists, 36; ideology and sex differences, 122–23; sociobiology, 124–25; ideology and premenstrual syndrome research, 205
Disease: observational bias and menstruation, 192
Dissection: anatomy and reproductive theory, 161
Domination: cell biology and politics, 181–82; ideology and premenstrual syndrome research, 205
Dualism: feminist theory of science, 10; gender, 109
Dysmennorhea: premenstrual syndrome research, 193

Ecology: Ellen Swallow, 4; Rachel Carson, 70–71; ecosystems and gynocentric science, 72
Economics: feminist science, 55–56; subject and non-neutrality, 63; sex differences, 122–23; language and objectivity, 126; Paracelsus

Economics—*continued*
and political dissent, 134; sociobiology and cell theory, 181; premenstrual syndrome, 188
Education: society and science, 120, 123
Edwards, Thomas: religious toleration and sexual equality, 139
Embryo: preformation doctrine, 163–68
Emotion: premenstrual syndrome research, 202–203, 206n
Empiricism: feminist epistemology, 25
Employment: status of women in science, 5; sex differences and ideology, 122–23
England: gender politics and philosophy of science, 135–40; law and premenstrual syndrome, 188
Epistemology: feminist research, 23–25; Copernican revolution, 64; truth and theory-choice, 95–96
Eroticism: feminist concept of science, 71
Eskimos: brain size and intelligence, 211, 212, 215
Evolution: feminist inquiry, 28; sociobiology and sex differences, 124
Experience: feminist research, 24, 29; women's as a scientific resource, 27–28; female social science, 96–98; male and female thought processes, 223–24

Facts: values, 100n; social enterprise, 119–20
Family: Blacks and feminist inquiry, 28; image and genetic theory, 179–82
Federation of Feminist Women's Health Centers: women and science, 129
Femininity: passivity, 184n
Feminism: theory of science, 9–11; gender, 26–27; method and science, 30; Barbara McClintock, 37; intentionality and behavior, 53; social sciences, 85–86; bias and values, 89; science and politics, 129; sex differences and thought processes, 222
Feminist science: feminine, 36, 37; Barbara McClintock, 39; scientific authority, 40–41; interactionism, 46–47, 55; institutional obstacles, 55–56; historical existence, 69–70
Fetus: Aristotle and reproductive theory, 150–51, 152; preformation doctrine, 163–68
Food science: gynocentric science, 72
Franklin, Rosalind: history of science, 4
Freud, Sigmund: subject and discourse, 67; women and sex difference, 123; femininity and passivity, 184n

Galen: sexist bias and reproductive theory, 152, 153–54, 158, 160, 163, 168
Garden, George: animalculist theory, 167
Geddes, Sir Patrick: model of sex determination, 173–74
Gender: social and biological construction, 8; feminist theory of science, 9–11; feminist research, 26–27, 29; race and class, 30n; sex, 33, 34, 35, 38, 39; Barbara McClintock, 37, 38; politics and science, 42–43; power and science, 43–44; language and science, 60; Keller's ontology, 109, 114; Boyle's philosophy and politics, 133, 134–40; bias and biology, 169, 184; cell biology, 181–82; organic chemistry and fertilization metaphors, 182–83; observational language and premenstrual syndrome research, 196; ideology and premenstrual syndrome research, 200, 205–206
Genetic engineering: politics and science, 140–41, 180
Genetics: sex determination, 178; sexualization of the cell, 179–82
Genitals: Galen and reproductive theory, 154
God: epistemology and science, 58; gender, 60–61
Gossip: gynocentric science, 73
Gough, William: women and subjection, 137
Green, Catherine: history of science, 4
Gregory of Nyssa, Saint: sexism and reproductive theory, 158

Hamm, Louis Dominicus: preformation doctrine, 164–65
Hartlib, Samuel: natural magic tradition and Boyle, 135
Hartsoeker, Nicolaus: preformation doctrine, 165
Health: pregnancy and childbirth, 77; sex differences and discrimination, 122–23; women's movement and science, 129
Heat: reproductive theory, 147–49, 152, 153–54
Heisenberg's uncertainty principle: physics and context, 127
Hemispheric specialization: intelligence studies and sexism, 212–13; cultural and sociological factors, 216; ideology and characterizations of distinction, 217–20; sex difference and maturation, 224n. *See also* Brain; Intelligence
Hermes Trismegistus: natural magic tradition, 134
Hierarchy: organization of knowledge, 75; theory and practice, 107; Boyle's theories, 143–44; genetic theory, 181
History: Renaissance, 21, 28; gynocentric science, 70; Dewey, 114; sex differences and discrimination, 123; gender politics and philosophy of science, 134–40
History of science: feminists and science, 4–5; feminist theory of science, 10; scientific method, 18; women scientists, 23–24, 74;

History of science—*continued*
Boyle and values, 49; gender and scientific theory, 133
Holistic processing: hemispheric specialization studies, 218–19
Home: women and knowledge, 128
Home economics: history of science, 4, 9; gynocentric science, 73
Homemaking: gynocentric science, 71
Homosexuality: biological determinism, 7
Hooke, Robert: gender politics and philosophy of science, 140
Hormones: sex and behavior, 51–52, 124
Hospitals: childbirth, 77–79
Human sciences: subjectivity and language, 58–59
Hylozooism: natural magic tradition, 134; Boyle, 136–37; sexual equality, 144
Hypotheses: feminist inquiry, 28; theoretical and observational language, 48, 49

Ideology: politics and science, 99; premenstrual syndrome research, 200–202; science and discrimination, 205–206; hemispheric specialization studies, 217–20
Imperialism: science and language, 59
Industry: feminist science, 56; ideology and discrimination, 122–23; premenstrual syndrome, 188
Inquiry: difference, 108; Dewey, 110–11; Keller, 111–14
Intelligence: science and status of women, 6. *See also* Brain; Craniometry; Hemisphere specialization; Lateralization
Intentionality: human behavior, 52, 53
Interactionism: feminist science, 46–47, 55; economics, 56n
Interconnection: feminist epistemology, 71
Intuition: science, 61; language, 62
Isomorphism: intuition and science, 61

Judaism: primacy of male generative powers, 158
Just, E. E.: genetic theory and gender politics, 180, 181

Katabolism: models of sex determination, 173, 174
Keller, Evelyn Fox: epistemological tradition, 104; ontology, 105–106, 108–10; inquiry, 111–14; gender and science, 168
Knowledge: power structure, 42; culture and choice, 54; gynocentric science, 72; hierarchical organization, 75; discourse, 92; Dewey, 106–107; Keller, 113

Labor: society and science, 121
Laboratory: social structure and context of science, 127
de Laguna, Andres: anatomy and reproductive theory, 160–61
Language: gender and science, 44; theoretical and observational, 48; passive voice, 51; science and the subjective, 58–59; intuition and science, 61–62; differently sexed, 64–65; women's, 65–66; maternal and paternal, 66–67; androcentrism and science, 81–82; objectivity and scientific method, 125–27; observational bias and premenstrual syndrome, 190–97; lateralization studies and sexism, 214
Lateralization: criticism, 7; intelligence studies and sexism, 213–16. *See also* Brain; Intelligence
Ledermuller, Martin Frobenius: animalculist theory, 167
van Leeuwenhoek, Anthony: preformation doctrine, 165; gender-related metaphor, 185n
Levellers: women and political dissent, 137–39
Library: science and values, 48
Lilbourne, John: sexual equality, 137
Linguistics: subject and non-neutrality, 63
Literacy: women's history, 75

McClintock, Barbara: feminist science, 9–10, 35, 36–39; Nobel Prize, 39–40, 41; holistic approach, 70; Keller's ontology, 108, 109–10
McClung, C. E.: models of sex determination, 174
Magic: gynocentric science, 72; political dissent, 134. *See also* Natural magic tradition; Witchcraft
Malpighi, Marcello: preformation doctrine, 164
Marriage: image and genetic theory, 179–82
Marxism: feminist method, 19; feminist scholars, 142
Masculinity: science and exclusion of women, 3, 8, 42; construction of science, 38; God in Western culture, 60–61
Massa, Niccolo: anatomy and sexist bias, 161, 163
Mathematics: subject and non-neutrality, 63; hemispheric specialization studies, 218
Menstruation: Aristotle and reproductive theory, 148–49, 159; cyclicity and premenstrual syndrome, 189; observational bias and negativity, 191; accident rates, 193; psychogenic factors, 202; social cognitive factors, 203
Metabolism: models of sex determination, 173–74
Methods: fetishization, 18; feminist alternatives,

Methods—*continued*
18–20; methodology, 21–22; feminist criticism, 183. *See also* Scientific method
Midwifery: definition of science, 9; gynocentric science, 71, 72; resurgence, 73; oral tradition, 74–75; obstetrics, 79–81
Military: feminist science, 56; gender-related metaphor, 176
Misogyny: science and ideology, 205
Model: choice and values, 52–53; intuition and science, 61; Freud and libido, 63; natural science and social change, 141–44; premenstrual syndrome research, 198–99
Moos Menstrual Distress Questionnaire: premenstrual syndrome and observational bias, 191
Moral development: androcentric assumptions, 18; methods and methodology, 21; feminist inquiry, 28; thought processes and sexual difference, 224
Morgan, T. H.: sexualization of the cell, 179–80
Mortality: childbirth, 77; cesarean section, 78–79
Mozans, H. J.: history of science, 4
Music: socialization and hemispheric dominance, 216
Mutation: Aristotle and reproductive theory, 152–53
Myth: reproductive theory and the sperm, 174–76, 184–85n; cellular and molecular biology, 179

National Mental Health Association: premenstrual syndrome, 189
Natural magic tradition: political dissent and philosophy of science, 134–40
Nature: science, 34, 35, 41–42; reality, 43; physics, 63–64; objectivity and scientific method, 125–26; Campanella, 134–35; Aristotle and reproductive theory, 152
Network model: natural science and social change, 142–44
Neurobiology: brain and behavior, 52–53
Nobel Prize: Barbara McClintock, 36, 39, 41
Nucleus: sexualization of the cell, 179–82

Objectivity: science and culture, xi; science and exclusion of women, 3; gender and science, 10; contextual values, 50; theory-choice, 95; inquiry relationship, 112; feminist and scientific method, 125; language, 125–27; natural science models, 141
Obstetrics: definition as science, 9; midwifery, 79–81
Occupational health: sex differences and discrimination, 122–23
Ontology: John Dewey, 105–108; Evelyn Fox Keller, 105–106, 108–10
Oral tradition: gynocentric science, 74–75; women's knowledge, 128
Order: Keller's ontology, 108–109, 110
Organic chemistry: gender-related metaphors, 182–83
Ovaries: Galen and reproductive theory, 154, 158; Vesalius and anatomy, 161; van Leeuwenhoek, 165
Ovism: preformation doctrine, 164, 165, 166
Ovum: sperm myth, 175–76, 184–85n; revisionism and reproductive theory, 176–79

Paracelsus: gynocentric science, 72–73; natural magic tradition and political dissent, 134
Passivity: science and language, 51, 125–26; sexism and reproductive theory, 151, 158, 159, 176, 184–85n; sex determination, 173–74, 178; cell biology, 181; gender-related metaphor and organic chemistry, 183
Patriarchy: female body, 202
Pharmacology: gynocentric science, 72–73
Phenomenology: feminist method, 19
Philosophy of science: feminist theory of science, 10; epistemology, 23; scientific method, 25; choice of methods, 27; realist tradition, 51; positivism, 88, 99; Boyle's biography, 133; non-scientific considerations, 141
Physics: theoretical and observational language, 48–49; subject and non-neutrality, 63–64; philosophy of science and positivist model, 88; context, 127; changes in interpretation, 183–84
de Pizan, Christine: history of science, 4
Policy: social science research, 27–28
Politics: science and women, viii, 70; empiricism, 24; duality and universality, 39–40; Boyle's theories, 49; choice of models, 53; liberation and feminist social science, 86, 87; theory-choice, 92, 99; women's experience, 96; science and context, 128–29
Positivism: theory-choice, 88, 89, 90, 99
Post-feminism: gender and science, 34
Post-modernism: feminist epistemologies, 25; gender and science, 34; difference, 43
Power: gender and science, 43–44
Practice: Dewey's ontology, 107–108
Pregnancy: midwifery and obstetrics, 80–81
Premenstrual syndrome: deviant behavior, 188–89; observational language and bias, 190–97; bias and explanatory models, 198–204
Problematics: scientific method, 28
Prozoology: gender distinctions, 182
Psychology: scientific method, 18; behaviorists, 27; brain and behavior, 53; non-neutrality

Psychology—*continued*
and subject, 63; subject and discourse, 67; schizophrenia, 127
Puberty: Aristotle and reproductive theory, 153
Puritanism: gender politics and philosophy of science, 134, 135; women and subjection, 137

Racism: feminist inquiry, 29
Ranelagh, Lady Katherine: Robert Boyle, 132
Realist tradition: fixed relations, 51
Reality: science, 42; truth and theory-choice, 93–94, 95–98
Relational processing: hemispheric specialization studies, 218
Relativism: theory-choice, 94, 95–96, 98–99; subjectivism, 100n; coresponsible option, 104–105
Renaissance: women's history, 21; feminist inquiry, 28
Reproduction: science and language, 60; sex differences and discrimination, 122–23; sociobiology, 124–25; Aristotle and sexist bias, 147–53, 158, 159, 168, 172, 173, 174; classical theorists, 153; Galen, 153–54, 158; primacy of male generative powers, 158–61, 163; preformation doctrine, 163–68; male bias in Western culture, 168–69; sex determination, 173–74, 177–78; sperm and myth, 174–76; revisionism and ovum, 176–79
Research: feminist inquiry, 26–29; morals and politics, 29–30
Robinson, John: sexual equality, 137

Schizophrenia: language and objectivity, 126–27
Scholarship: feminism and science, 3; history of science, 5; language, 82
Science: politics and women, viii, 42; gender, x–xi, 38; male domination, 3; feminine, 8–9, 36, 37; feminist theory, 9–11; meaning and science studies, 33; nature, 34, 35, 41–42; authority, 40–41; difference and duality, 42; gender and power, 43–44; use of term, 45, 81–82; subjectivity and language, 58–59; as political term, 70; androcentric model, 71–72; title and institutions, 73–74; as social enterprise, 119–21; social context, 140–41; creativity, 183; ideological assumptions, 199; ideology and discrimination, 205–206, 217; hemispheric specialization studies and discrimination, 213. *See also* Feminist science; History of science; Philosophy of science
Scientific method: feminist method, 20, 125–26; bias, 23; philosophy of science, 25–26; problematics, 28. *See also* Methods
Sectarianism: gender politics and philosophy of science, 135–40
Semen: sexist bias and reproductive theory, 148–50, 151, 159, 160; preformation doctrine, 164–65
Sex: gender, 33, 34, 35, 38, 39; reality, 43; hormones and behavior, 51–52; ideology and reality of difference, 121–22; research, 123–24; determination theory, 173–74, 177–78, 185n; differences and lateralization studies, 214–15; difference and modes of thought processing, 220
Sexism: androcentrism, 145n; women and hemispheric specialization research, 224n
Social sciences: methods and androcentric assumptions, 17, 18; feminist method, 20, 30; feminist research, 21–22; empiricism and methodology, 23; feminist epistemologies, 23–24; gender, 27; politics and research, 27–28; status as science, 73; relationship to feminism, 85–86; androcentric theories, 87–88
Society: status of women, 6–7; conditions and kind of science, 46; facts and science, 119–21; context of science, 140–41
Sociobiology: status of women, 6; sex differences, 124–25; cell theory and hierarchy, 181; feminist criticism, 183
Sociology: feminist method, 20; methods and methodology, 21; understanding of epistemologies, 24; laboratory and context, 127
Sperm: myth and reproductive theory, 174–76, 184–85n; re-evaluation of reproductive theory, 177; sexualization of the cell, 179
Split-brain theory. *See* Lateralization
Statistics: premenstrual syndrome research, 193, 195
Steady state: genetic theory and politics, 180, 181
Stereotype: femininity and McClintock, 38
Subjectivity: language of science, 58
Swallow, Ellen: history of science, 4, 8–9
Swammerdam, Jan: preformation doctrine, 163–64
Symptoms: premenstrual syndrome research, 194–95

Teaching: feminists and science, 4, 10
Technology: science and society, 121; politics, 128
Textbooks: sperm myth and metaphor, 175, 185n; sexism and reproductive theory, 176; genetic theory and politics, 181; organic chemistry and fertilization metaphors, 182; gender-related images, 185n

Theology: natural magic tradition and political dissent, 134–40; primacy of male generative powers, 158
Theory: data, 48, 49; Dewey's ontology, 107
Theory-choice: feminist social science, 86, 87; positivist conception, 88, 89; Holistic Model, 90–92, 93–94; Constructivist Model, 92–93, 94; women's experience, 96–98; relativism, 98–99
Thermodynamics: Freudian psychology, 63
Thinking: processing strategies, 220–21; society and culture, 222–23. *See also* Brain; Hemispheric specialization; Intelligence; Lateralization
Thomas Aquinas, Saint: sexism and reproductive theory, 158, 172
Thomson, J. Arthur: model of sex determination, 173–74
Truth: science and authority, 40–41; Holistic Model of theory-choice, 93–94; Constructivist Model of theory-choice, 94–96; women's experience and theory-choice, 96–97; semantic and correspondence theories, 101n

Universality: intuition and science, 61. *See also* Difference; Dualism

Values: science and bias, 7–8; constitutive and contextual, 47–50; choice of models, 52–53; positivist conception of theory-choice, 88–89; models of theory-choice, 90, 92, 93; facts, 100n; Aristotle and reproductive theory, 150–51
Vesalius, Andreas: anatomy and dissection, 161
Violence: suppression of gynocentric traditions, 75; sperm myth and gender-related metaphor, 176, 185n; premenstrual syndrome and observational bias, 191

Waddington, C. H.: sexualization of the cell, 180
Whitney, Eli: history of science, 4
Winstanley, Digger Gerrard: natural magic tradition and sectarianism, 135, 136
Witchcraft: midwifery, 73; suppression of gynocentric traditions, 75
Women: as scientists, 35–36; traditional view of science, 41; diversity of experience, 47; language, 65–66; ideology and reality, 121–22; roles in science, 127–28; health movement and science, 129; status in seventeenth-century England, 137; Levellers and political dissent, 137–38; hemispheric specialization research, 224n
Work: women and science, 74; language, 82